From Semiclassical Semiconductors to Novel Spintronic Devices

Editor

Halyna Khlyap

Kaiserslautern
Germany

CONTENTS

FOREWORD

The electronic book (e-lectures) "From Semiclassical Semiconductors to Novel Spintronics Devices" edited by Dr. H. Khlyap presents an interesting supplement to traditional manuals and textbooks devoted to microelectronics and physics of semiconductor devices.

E-lectures provide very useful information about narrow-gap semiconductor solid solutions Hg(Zn)CdTe (there is a material of choice for novel IR-photodevices), thermodynamical description of high-effective solar energy conversion, basic knowledge related to the semiconductor devices modeling and a good review of novel spintronics devices. This supplement can be significantly helpful for undergraduate, graduate and early PhD students beginning their professional activities in modern semiconductor physics, nano(micro)electronics and technologies.

Supplying necessary fundamental information, this collection of e-lectures may also serve as a reference book for professionals switching out their branches of research and development.

V. Bilozertseva
National Polytechnical University "KPI"
Kharkiv
Ukraine

PREFACE

The dramatic developments in the area of semiconductor device technology and nanotechnologies over the last decades has placed increasing demands on the fabrication and design technologies, where it is now necessary to implement nanoscale gate geometries to achieve the highest performance. The need to understand and model the operation of nano-devices is fundamental to future development and optimal design.

The key to the understanding of semiconductor nanoscaled active elements and devices based on smart materials (for example, ZnCdHgTe for far-infrared optoelectronics) lies in development of sophisticated models representing the physical and electrical characteristics as well as operation conditions.

Conventional silicon-based solar cells production accounts principal characteristics of the semiconductor compounds presenting the main building part of the device. The first part of the lecture devoted to high-effective solar cells reports unique results on fabrication and quite uncommon performance of Si-based solar cells with novel contacts chemically deposited on traditionally prepared semiconductor part of the solar cell. How does the high-effective solar radiation conversion occur and what thermodynamics aspects are to be involved in theoretical explanation and further practical applications? The next part of the e-lecture proposes answers. The thermodynamics is an effective tool used in the physics of the Universe. Photonic condensate and the relic radiation are not only exotic objects for theoretical astrophysics but also present possible power tools for achieving high-effective energy conversion. This question is considered in further parts of this lecture.

Spintronics, or spin electronics, involves the study of active control and manipulation of spin degrees of freedom in solid-state systems. The basic physical principles underlying the generation of carrier spin polarization, spin dynamics, and spin polarized transport in semiconductors and metals are of special importance. Spin transport differs from charge transport in that spin is a nonconserved quantity in solids due to spin-orbit and hyperfine coupling [3].

This eBook is intended to provide review lectures on latest achievements in growth technology, characterization methods and nanoscaled active elements modeling for novel device design and applications for graduate and PhD students as well as for professionals in mentioned fields of device technology and modeling.

Halyna Khlyap
Kaiserslautern
Germany

List of Contributors

Viktor I. Laptev

Russian New University, Moscow, Russian Federation, and Holzgerlingen, Germany.

Petro G. Sydorchuk

State Pedagogical University, Drohobych, Ukraine.

Halyna Khlyap

Kaiserslautern, Germany.

Send Orders of Reprints at reprints@benthamscience.net

CHAPTER 1

Narrow Gap Semiconductors Based on Mercury-Cadmium Telluride

Petro G. Sydorchuk and Halyna Khlyap[*]

State Pedagogical University, 24 Franko str., UA-82100 Drohobych, Ukraine

Abstract: The fundamentals of technologies applied for preparation of narrow-gap semiconductors Hg(Zn)CdTe are presented. The lecture discusses questions related to the technologies and device applications of active elements based on HgZnCdTe materials.

Keywords: Cadmium mercury telluride, quaternary solid solutions, phase diagrams, infrared detectors, current-voltage characteristic, capacitance-voltage characteristic, energy band diagram, atmospheric oxygen.

1. INTRODUCTION

Narrow gap semiconductors (E_g is about 0.18 eV at 300 K) are the most attractive materials widely used for practical applications in micro-and optoelectronics, especially for fabrication of infrared (IR) photodetectors and photodiodes.

In particular, the electron energy E in these materials is comparable with the value of forbidden gap E_g. The conduction band is characterized by the small effective mass and high mobility of the charge carriers as well as by low density of states which is proportional to $(m^*)^{3/2}$. The Fermi level is placed in the conduction band even at relatively low concentration of charge carriers. As the energy of interaction between the valence band and the condution band is also comparable with E_g the dispersion law for the charge carriers in the conduction band departs from the semi-classical quadratic function. The flexibility of the band structure gives a wide range of practical applications in various fields of opto-and microelectronics.

2. SOLID SOLUTIONS OF BINARY COMPOUNDS HgTe AND CdTe

In 1959 Lowson and co-authors [1] have shown for the first time that the alloys of

---

*Address correspondence to Halyna Khylap: State Pedagogical University, 24 Franko str., UA-82100 Drohobych, Ukraine; Tel: +49 631 414 4865; Email: gkhlyap17@yandex.ru

HgTe and CdTe are very perspective materials for preparation of intrinsic photodetectors operating in IR-spectral range. Since earlier 60th the solid solutions of cadmium telluride, mercury telluride and zinc telluride (HgTe-CdTe, HgTe-ZnTe) are attracting attention of materials scientists.

Thanking to a large number of investigations it had been found that the mercury telluride (HgTe) is characterized by the inversed band structure. The energy bands Γ_8 degenerated at x = 0 are placed above the band Γ_6. Thus, the parameter of the Kane model $E_g = E(\Gamma_6)-E(\Gamma_8)$ is negative for the mercury telluride. Unlike to the classical semiconductors, the value of E_g does not coincide with the forbidden gap in these materials. In this case the forbidden gap is zero due to the degeneration of the valence band and the conduction band which are belonging to the same point Γ_8 of the band structure. At the same time, the dispersion laws of the conduction band Γ_8 and the valence band for the light holes Γ_6 are principally determined by the Cane theory (as in the classical semiconductors).

Cadmium telluride (CdTe) is a direct gap semiconductor with E_g = 1.45 eV at 300 K. Both binary compounds are crystallized in the zinc blende structure forming the continous row of substitutional solid solutions. Under formation of solid solutions the band Γ_8 can be presented as a fixed one, and re-building the gap structure under the change of the composition x of solid solution may be considered as a result of the uniform motion of the Γ_6 band relative to the Γ_8 band as the composition x changes, so it is possible to obtain semi-metal and semi-conducting compounds separated by the special value of x corresponding to the gapless semiconductor.

Investigations of luminescence, photoconductivity and optical absorption have demonstrated that the gap value has almost linear dependence on the composition of the solid solution. These experimental data made it possible to derive empirical expressions for the function E_g = f(x, T). For example [2], this function describes the parameters for a solid solution $Cd_xHg_{1-x}Te$ (0.13 < x < 0.60) at 20K < T < 300K as follows:

$$E_g = -0,25 + 1,59x + 5,233 \cdot 10^{-4}(1 - 2,08x)T + 0,327x^3$$

Fig. (**1**) plots the forbidden gap for of mercury-cadmium telluride and mercury-zinc telluride solid solutions as a function of their composition x. As is seen, the energy distance $E_g = E(\Gamma_6) - E(\Gamma_8)$ is negative in HgTe, so that we have semi-metallic material for all values of x corresponding to $E_g < 0$. For the composition x corresponding to $E_g > 0$ we have a semi-conducting compound with values of the gap up to $E_g(CdTe)$ or $E_g(ZnTe)$. If $E_g = 0$ (with a respective composition) at a given temperature we have a gapless semiconductor. In this case the effective mass of the charge carriers at the bottom of the conductance band is zero and their mobility is constant. One should note that the energy has a quasi-linear dependence on the wave vector [3].

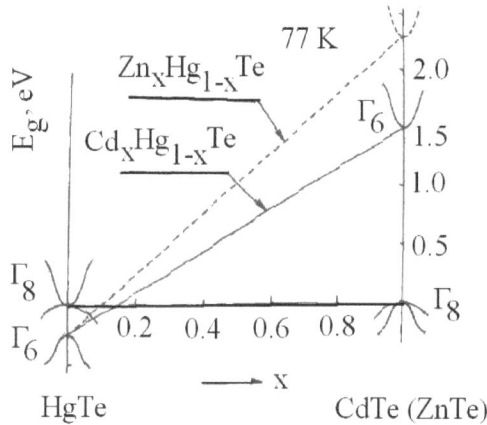

Figure 1: The gap structure at 77 K as a function of composition x for solid solutions $Cd_xHg_{1-x}Te$ and $Zn_xHg_{1-x}Te$.

The electro-physical properties of solid solutions $Cd_xHg_{1-x}Te$ are good investigated [4, 5]. The intrinsic conductivity is observed in the wider temperature interval in comparison with other semiconductors. The electron mobility may be in a few orders of magnitude larger than that of holes. Under these conditions even the material of a p-type conductivity (the holes are the major charge carriers) demonstrates properties of the electronic semiconductor at low temperatures (the Hall coefficient is negative). Some peculiarities of the temperature dependence of the Hall coefficient observed for HgCdTe material are caused by choosing the composition or doping inhomogeneities appeared under the technological processes.

The basic defects in HgCdTe solid solutions are the mercury or tellurium vacancies and the interstitial atoms of Hg. This experimental fact opens new ways for changing the type of conductivity by the heat treatment of the samples at different pressures of the mercury saturation vapor (Hg is the most volatile component of the solid solution). It is found that the vacancies of mercury lead to appearance of the acceptors; the vacancies of tellurium and the interstitial mercury give rise to the donors. The most important mechanisms of the charge carriers scattering in the HgCdTe solid solutions are the scattering by the charged centers (at low tempeatures), the polarization of the optical phonons play a sufficient role at elevated temperatures.

Thus, the unique features of the HgCdTe solid solutions make this semiconductor a very important material for producing IR-devices.

Solid solutions $Zn_xHg_{1-x}Te$ have also attracted attention of researchers [6, 7]. It was found out that the structure and physical properties of this material are similar to those of $Hg_{1-x}Cd_xTe$. However, the technology of $Zn_xHg_{1-x}Te$ solid solution is more complicated due to considerable difference of the lattice constants for binary compounds HgTe and ZnTe (about 5-6%, Fig. (**2**)).

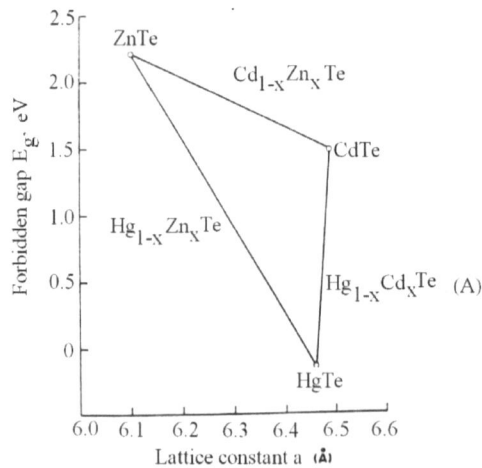

Figure 2: The diagram *gap-lattice constant* for II-VI ternary alloys at 300 K.

Authors [8] have theoretically shown that the weak *Hg-Te* bond is destabilized under interaction with *CdTe*, but it is stabilized by *ZnTe*. Results of these calculations have stimulated the development of technological processes, in particular, the liquid phase epitaxy from Te-rich solution which had been earlier proposed for preparation of solid solutions $Hg_{1-x}Cd_xTe$. The epilayers $Hg_{1-x}Cd_xTe$ were grown on monocrystalline substrates $Zn_xCd_{1-x}Te$ with a low Zn content (~1%) for a better coinsidence of the lattice constants of the substrate and epilayer (in case of using CdTe as a substrate the discrepancy between these parameters had been a challenge for obtaining epilayers of a device quality).

As the next step the functions $E_g = f(x)$ and $a = f(x)$ (*i.e.* how do the forbidden gap and the lattice constant of epilayers depend on the composition x) were examined. The technique of defining the bulk crystal density, the electron-probe microanalysis as well as the X-ray analysis with using the Vegard's law were applied to study of the $Zn_xHg_{1-x}Te$ solid solution. At the room temperature the function $a = f(x)$ is:

$$a(x) = 0,6463-0,360 \, x \, (nm) \tag{1.1}$$

Dependence E_g on the composition and temperature is empirical; the most usable expression [9] is as follows:

$$E_g(x,T)=-0.3+(3.24\cdot10^{-2}x^{1/2})+(2.731x-0.629x^2)+0.533x^3+(5.30\cdot10^{-4}T(1-0.76x^{1/2}-1.29x)) \tag{1.2}$$

As the bond length of the zinc telluride (0.2406 nm) is smaller than those of HgTe (0.2797 nm) and CdTe (0.2804 nm) the energy of dislocation formation and microhardness for solid solutions $Zn_xHg_{1-x}Te$ and $Zn_xCd_{1-x}Te$ are larger compared with $Cd_xHg_{1-x}Te$. Fig. (**3**) demonstrates the dependence of the Vickers-microhardness H on composition x for solid solutions $Cd_xHg_{1-x}Te$ and $Zn_xHg_{1-x}Te$.

Both curves have maximums at the larger values of x, the values of microhardness H for $Zn_xHg_{1-x}Te$ are considerably higher that those for $Cd_xHg_{1-x}Te$. This fact points out the better mechanical properties and the more perfect structure of solid solutions $Zn_xHg_{1-x}Te$ in comparison with $Cd_xHg_{1-x}Te$. This is an advantage under fabrication of photodetectors.

Investigations of electron effective mass as a function of the composition of solid solution $Zn_xHg_{1-x}Te$ showed that the value of m* corresponding to the composition under which the transition *semimetal-semiconductor* occurs, is equal to zero, the same fact is observed for $Cd_xHg_{1-x}Te$ (Fig. (**4**)).

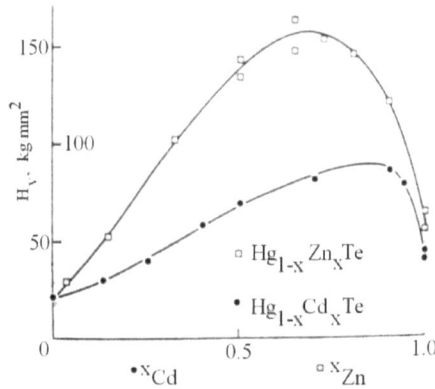

Figure 3: Dependence of the microhardness H on the composition x for solid solutions $Cd_xHg_{1-x}Te$ and $Zn_xHg_{1-x}Te$.

The valence band of this semiconductor ($Zn_xHg_{1-x}Te$) is described as a simple parabolic band with the effective mass of heavy holes $m_h^* = 0.60\ m$. The effect of the light holes band on the transport processes is negligible.

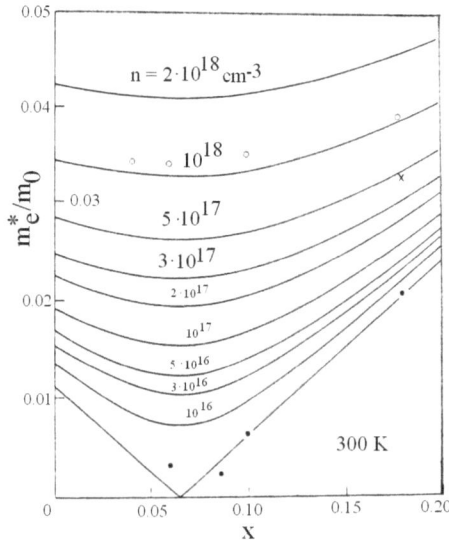

Figure 4: The electron effective mass as a function of the composition x for solid solutions $Zn_xHg_{1-x}Te$ calculated for different carrier concentrations, and the experimental values of m^* at the bottom of the conduction band (dark circles) and for samples with electron concentration $n = 10^{18} cm^{-3}$, m_0 is a mass of a free electron [10].

3. QUATERNARY SOLID SOLUTIONS $Zn_XCd_YHg_{1-X-Y}Te$

Quaternay solid solutions $Zn_xCd_yHg_{1-x-y}Te$ were obtained for the first time in 1979 [11]. This material is considered as a perspective alternative to the well-known narrow-gap semiconductor $Cd_xHg_{1-x}Te$ as a basis for designing novel devices for IR-optoelectronics. Authors [8] have showed that embedding Zn atoms in the lattice of $Cd_xHg_{1-x}Te$ provides stabilization of weak bonds in the lattice and increase of the defect formation energy resulting in more perfect structure of the crystals and enhanced mechanical toughness. The investigations of the microhardness performed for quaternary solid solutions $Zn_xCd_yHg_{1-x-y}Te$ [12], (Fig. (5)) have confirmed these predictions.

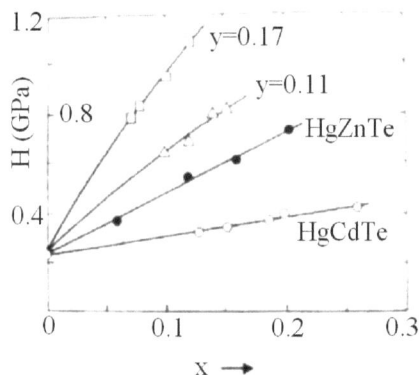

Figure 5: Vickers microhardness of $Zn_xCd_yHg_{1-x-y}Te$ as a function of composition.

As is shown, the microhardness of the quaternary solid solution $Zn_xCd_yHg_{1-x-y}Te$ at small values of x and y is sufficiently higher than that for ternary solid solutions $Cd_xHg_{1-x}Te$ and $Zn_xHg_{1-x}Te$.

Examinations of the Hall effect and electroconductivity of this new material demonstrated a good quality of the samples $Zn_xCd_yHg_{1-x-y}Te$. Electrical properties of the compound are better or they have the same quality as those of the $Cd_xHg_{1-x}Te$. Fig. (6) plots temperature dependences of the Hall coefficient and electroconductivity for the sample $Zn_xCd_yHg_{1-x-y}Te$ ($x = 0,12$; $y = 0,10$) prepared by the liquid phase epitaxy fromTe-rich solutions. The intrinsic conductivity is observed in the temperature range (300-170) K; at lower temperatures the impurity conductivity takes place. The electron concentration of narrow gap samples as estimated to be about 10^{15} cm^{-3}, the mobility of charge carriers (electrons) had reached 10^5 cm^2/V$\cdot$s. The charge carriers' concentration of $Zn_xCd_yHg_{1-x-y}Te$ can be changed by the thermotreatment of as-grown layers and monocrystals in saturated mercury vapor changing the namber of vacancies of the most volatile element. Magnetophonon resonance observed in this material had also showed a good quality and perfect structure of $Zn_xCd_yHg_{1-x-y}Te$ [13].

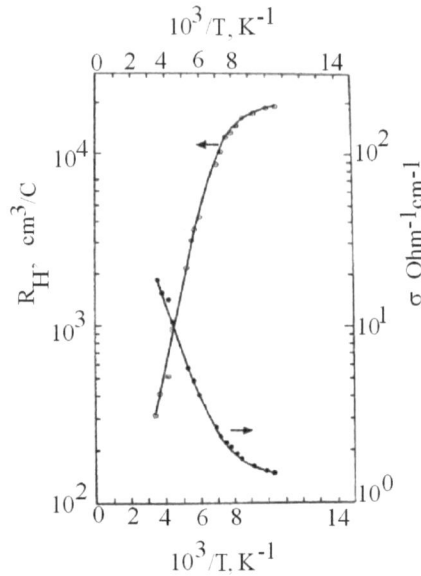

Figure 6: Temperature dependences of the Hall effect and the electroconductivity of the sample $Zn_xCd_yHg_{1-x-y}Te$ (x =0,12; y = 0,10).

Thin film technologies, in particular, pulse laser deposition technique, made it possible to prepare heterostructures $CdTe$-$Zn_xCd_yHg_{1-x-y}Te$ [14-16].

4. PHASE DIAGRAMS

Investigation of phase diagrams in geterogeneous systems is a necessary tool for preparation of multicomponent semiconductors. The diagrams *composition-melting* temperature and *composition-vapour pressure* are used for examination of phase equilibrium in binary systems. The process is more complicated for compounds containing volatile components. In this case the diagrams *vapour pressure-temperature* are used.

Fig. (**7**) plots the diagrams *temperature-composition (T-x)* for binary compounds *HgTe, CdTe* and *ZnTe*. They are similar. The system has components which are melting congruently, and the eutectic of the components with low-melting point is degenerated.

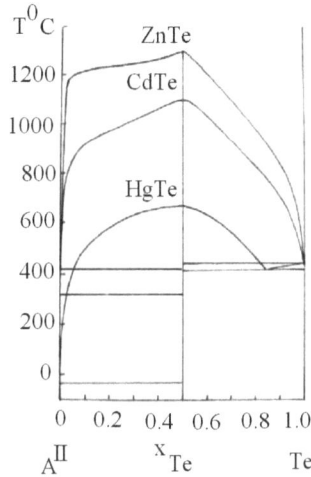

Figure 7: Temperature-composition diagrams for binary systems *Hg-Te, Cd-Te* and *Zn-Te*.

Fig. (**8**) illustrates ternary diagrams *(T-x)* built on base of the concentration triangles; the angles correspond to the components of the ternary solid solution, the ribs of the prism are corresponding to the temperature.

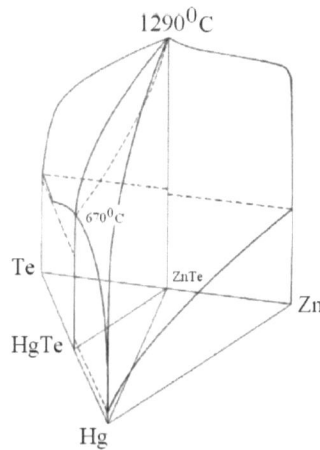

Figure 8: *T-x* diagram of phase equilibrium in the system *Hg-Zn-Te*.

It is necessary to consider poly-thermal cross-sections of the ternary system at the formation of binary compounds (the cross-section *HgTe-ZnTe* in Fig. (**8**)) as quasi-binary ones. At the same time, the binary compounds are considered as components of the investigated system. For example, the quasi-binary diagram temperature-composition of the system *HgTe-CdTe* is typical for the diagrams characterizing the continouos row of solid solutions (the type I according to the Rosebom classification) (Fig. (**9**)). The diagram for the sytems *HgTe-ZnTe* and *CdTe-ZnTe* are analogous.

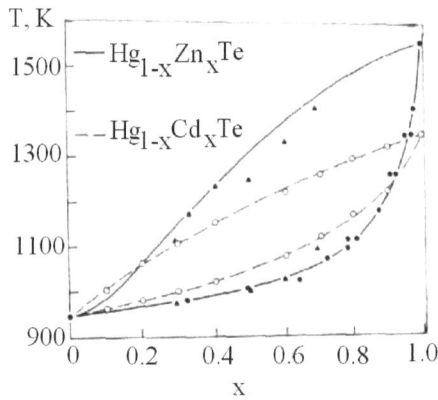

Figure 9: *T-x* diagrams of quasi-binary systems *HgTe-CdTe* and *HgTe-ZnTe*.

Fig. (**10**) plots the state diagrams for quaternary compounds. The panel *(a)* shows the system *Hg-Cd-Zn-Te*. The ribs of the tetrahedron are presented by the binary compounds forming by the each of three metals with tellurium. They contain the binary systems *HgTe, CdTe,* and *ZnTe*. The plains correspond to ternary systems *Hg-Cd-Te, Hg-Zn-Te, Cd-Zn-Te*. The plane *HgTe-CdTe-ZnTe* divides the tetrahedron *Hg Cd Zn Te* in two parts, the upper is of special interest: this is a second tetrahedron *HgTe-CdTe-ZnTe-Te*, which technologically describes the crystallization of solid solutions $Zn_x Cd_y Hg_{1-x-y} Te$.

As is seen from the diagram presented by the tetrahedron *HgTe-CdTe-ZnTe-Te*, quaternary solid solutions $Zn_x Cd_y Hg_{1-x-y} Te$ can be prepared by two technological processes: a) from the solutions of mercury, cadmium and zinc in Te; b) from the stoichiometric melts. In the first case the special point defining the composition of the liquid phase lies inside the tetrahedron *HgTe-CdTe-ZnTe-Te*, and in the case

b) point is placed in the second case in the plane *HgTe-CdTe-ZnTe*. If this point is inside the tetrahedron *HgTe-CdTe-ZnTe-Te,* for example, in the plane *abc*, it is necessary to determine in-advance projections of the isothermal sections of the liquidus-surface in the mercury angle of the system $(Zn_uCd_g Hg_{1-u-g})_{1-w}Te_w$ for a selected value *w* in order to prepare an epitaxial layer with predefined composition. Fig. (**10b**) shows these projections for *w = 0.80* with step 10°. The experimental results plotted in Fig. (**10b**) are obtained by authors [14].

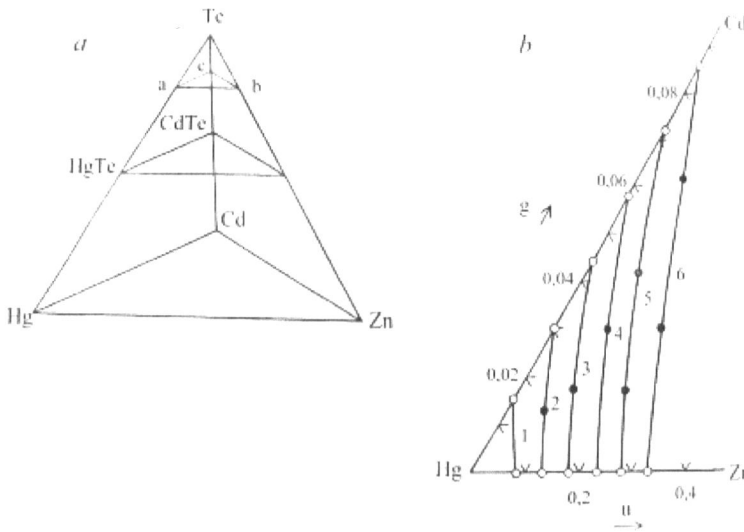

Figure 10: Phase diagram for a quaternary system системи *Hg-Cd-Zn-Te (a)* and liquidus isotherms in th mercury angle of the system $(Zn_uCd_gHg_{1-u-g})_{1-w}Te_w$ for *w= 0.80 (b)* and temperatures ($^{\circ}$C): 1-480; 2-490; 3-500; 4-510; 5-520; 6-530.

At the growth of epitaxial films $Zn_xCd_yHg_{1-x-y}Te$ from the stoichiometric melts one uses a quasi-ternary diagram temperature-composition (Fig. (**11**)) built on the base of known quasi-binary phase diagrams *HgTe, CdTe,* and *ZnTe*. Points T_1, T_2, and T_3 on the axes of temperature are the melting points of the binary compounds *HgTe, CdTe,* and *ZnTe*, respectively. Fig. (**11**) plots the isothermal section of the diagram for a quasi-ternary system *HgTe-CdTe-ZnTe*. The isotherm at the temperature $T\ (T_1 < T < T_2 < T_3\ )$ crosses the liquidus-surface along the curve A'B', and the solidus-surface along the curve C'D'. Projections of these curves to the plane of the concentration triangle produce curves AB and CD, respectively.

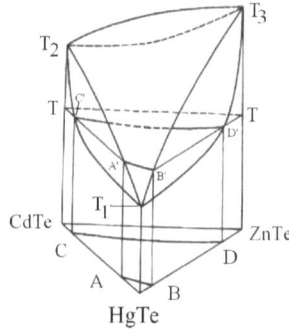

Figure 11: The diagram *temperature-composition* of quasi-ternary system *HgTe-CdTe-ZnTe*.

The narrow-gap quaternary solid solutions $Zn_xCd_yHg_{1-x-y}Te$ are of particular importance for IR-optoelectronics, so that the special points of composition for a liquid pahse with accounting segregation coefficients for zinc and cadmium are to be close to the angle HgTe of the quasi-ternary diagram (*T-x*) of the system *HgTe-CdTe-ZnTe* [14].

5. LIQUID PHASE EPITAXY AND LASER ABLATION TECHNIQUES FOR PREPARING NARROW-GAP SEMICONDUTORS

Liquid phase epitaxy (LPE) is one of the most successfully developed technologies for preparing semiconductor films. The word "epitaxy" comes from Greek words: *epi* = upon; *taxis*= ordered, and it means "growth of a crystal on a substrate with the same crystallographic structure as the substrate". The LPE as a technological method is simple, relatively inexpensive, and it is suitable for selective growth. As we have mentioned above, this technique is applied for the growth of the narrow-gap semiconductor mercury-cadmium-telluride solid solutions. In a complicated furnace (Fig. **(12)** [15]) the compound materials are molten up. Then the substrate is dipped into the melt. By cooling down with an adequate rate a thin single crystalline film begins to grow with the surface of the substrate as nucleation area [15]. The temperature during this process is around 465°C. After about 40 minutes the substrate is pulled out of the melt [15].

The details of technology are described in [16-22]. The authors have showed the methods and experimental setups for preparation of original monocrystals further used at fabricating epitaxial layers. Now we are going to the discussion of techniques for obtaining mercury cadmium telluride-based narrow-gap semiconductor thin films.

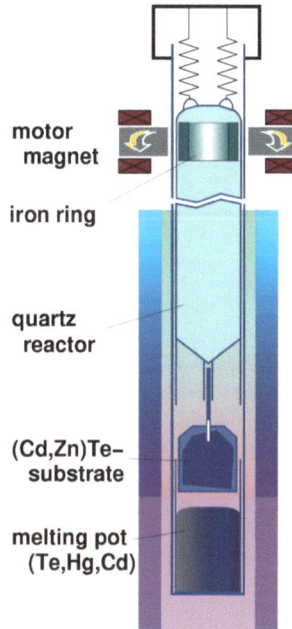

Figure 12: A schematic sketch of the furnace used for HgCdTe LPE growth [15].

LPE has significantly simplified the processes of prepairing devices for IR-optoelectronics. Previous technologies had used bulk ingots as origins for mechanical cutoff of the wafers. Such a method led to material losses and defect formation in the wafers considerably suppressed the performance of the active element. The epitaxial technology allows avoiding mechanical working-out of the proto-devices and gives much more possibilities to use the advantages of planar techniques.

The most usable methods applied for narrow-gap semiconductor solid solutions $Cd_xHg_{1-x}Te$ vapor phase epitaxy, liquid phase epitaxy, and molecular beam epitaxy [24, 25].

The layers of solid solutions $Zn_xHg_{1-x}Te$ [25] and $Zn_xCd_yHg_{1-x-y}Te$ [26] were prepared on the substrates $Zn_zCd_{1-z}Te$ from $HgTe$ origin. The growth was performed in two-zone furnace in the temperature range (600-640) $^{\circ}$C. The composition of the substrates $Zn_zCd_{1-z}Te$ was changed from $z = 0.05$ to 0.52. Mercury vapor pressure was in range $(1.0-5.0) \times 10^5$ Pa. Fig. (**13**) shows the concentration profile of the as-grown layers.

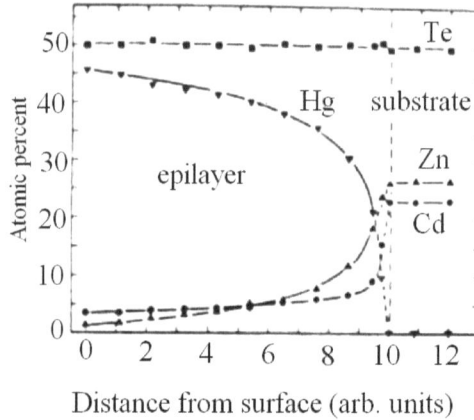

Figure 13: Composition profiles in $Zn_xCd_yHg_{1-x-y}Te$ epilayer grown on a $Cd_{0.48}Zn_{0.52}Te$ substrate at 600 °C under a mercury pressure of 1.5×10^5 Pa. A unit of the distance is equal to 5.5 μm.

Good results were obtained under epitaxial growth of $Cd_xHg_{1-x}Te$ layers on (111)-oriented $CdTe$ substrates from the stoichiometric alloys [30]. The process was carrid out in a closed system due to considerable pressure of the components (Fig. (**14a**)). The composition of the liquid phase for growing the layer of the solid solution $Cd_xHg_{1-x}Te$ with predefined composition was chosen from the $(T-x)$-diagram of quasi-binary system $HgTe-CdTe$ (Fig. (**9**)). The melt had been heated up to the temperature somewhat higher that the temperature of liquidus in order to achieve a complete homogenization of the technological melts. The melt had been contacted with the substrate to create over-saturation by forced cooling with 5^0 step.This technique allowed growing an epitaxial layer. The composition profiles of the films with thickness up to 150 μm were registered by the X-ray microanalysis (Fig. (**14b**)).

As is illustrated in Fig. (**14b**), the composition of the solid solution decreases with a distance d from the substrate from 1.0 to 0.2, and then it becomes constant. Analysis of these data demonstrated that near the substrate the experimental dependence $x(d)$ is qualitatively presented by the calculated curve 2 in assumption of the cadmium diffusion from the substrate into epilayer with diffusion coefficient $D_{Cd} = 10^{-9}$ cm^2/s. At the distances $d > 30$ μm this dependence coincides with the calculated function for a normal crystallization with the coefficient of segregation equal 2.8 (curve 1 in Fig. (**14b**)). As the time of epitaxial growth has increased, the section of variable composition in Fig. (**14b**) has also increased. The quality of the as-grown layers enabled to design multielement IR-photodetectors.

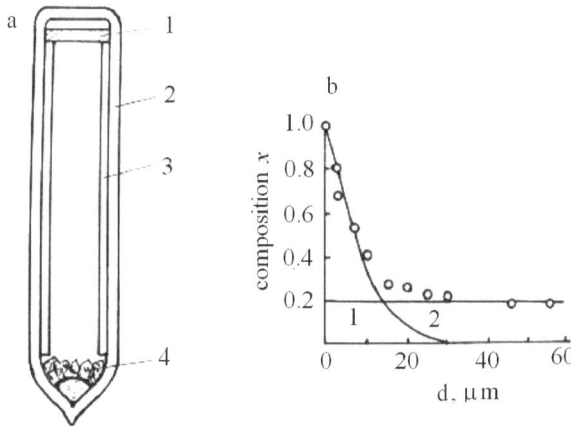

Figure 14: Quartz vessel for the epitaxy of $Cd_xHg_{1-x}Te$ solid solutions from the melt (a): 1-*CdTe* substrate; 2-quartz vessel; 3-quartz tube; 4-components for the growth of the solid solution; composition profiles in the as-grown layers (b).

Electrophysical investigations showed that layers grown on undoped CdTe substrate were of *p*-type conductivity. The concentration of the holes was estimated to be $5 \cdot 10^{16}$ cm^{-3}, their mobility μ was equal to 400 $cm^2/V \cdot s$. The films grown on CdTe:In wafers were of *n*-type conductivity with $n=2 \cdot 10^{15}$ cm^{-3} and $\mu = 1 \cdot 10^5$ $cm^2/V \cdot s$ [31].

Liquid phase epitaxy made it possible to obtain layers of quaternary solid solutions $Zn_xCd_y Hg_{1-x-y}Te$ from tellurium solutions on the substrates (111) *CdTe* or $Zn_xCd_{1-x}Te$ *(x<0.05)* of 18 mm diameter [13]. The content of the components had been chosen according to liquidus isotherms of the system $(Zn_uCd_gHg_{1-u-g})_{1-w}Te_w$ *(w=0.80)*. The mixture of growth components was previously synthesized as a tablet. The growth chamber made from graphite is shown in Fig. (**15**). This modified liquid phase epitaxy method gives a possibility to obtain layers with a surface free from tellurioum soluted in the technological solvent.

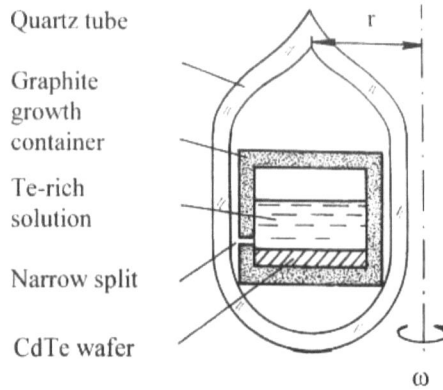

Figure 15: Schematic drawing of the growth system [15].

The modified method mentioned above allows to remove technological solvent-melt particles from the surface of the as-grown layer under action of the centrifugal force $f = m \, \omega^2 \, r$. Under rotation of the growth chamber placed at the distance r from the rotation axis with definitive angle velocity ω this force can be larger than the gravitation force, then the technological solution is completely removed.

Fig. (**16**) plots concentration profiles of $Zn_xCd_yHg_{1-x-y}Te$ films grown on (111)-oriented *CdTe* wafers from the teluriom solution (open circles) and from the stoichiometric melts (dark circles).

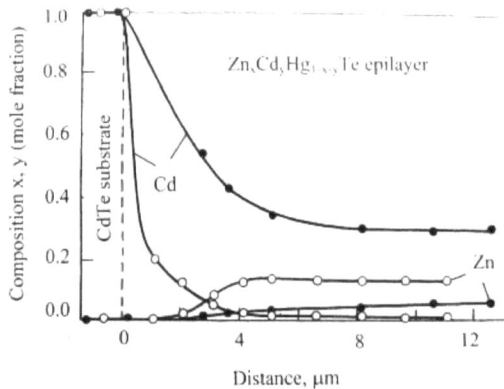

Figure 16: Composition profiles in $Zn_x \, Cd_y \, Hg_{1-x-y}Te$ epilayers grown on CdTe substrates from the solution of Zn, Cd and Hg in Te (hollow circles) and from stoichiometrric melts (solid circles) [13].

As is shown, the concentration profile sufficiently depends on the growth mode. Under epitaxy from the tellurium solution the range of the composition change near the interface layer-substrate is lower than 4 μm. In the layers obtained from the stoichiometric melts this range extends to 8 μm due to cadmium diffusion during the growth of the film.

Molecular beam epitaxy (MBE) as well as other technologies [32, 33] has also successfully applied for deposition of HgCdTe-based materials [34-36], for example, $Zn_xCd_{1-x}Te$ [36] layers were of device quality. The structures obtained by MBE technology allowed designing photosensitive active elements which can be irradiated from the side of wide gap component thus increasing the quantum efficiency and lowering the noise level.

Now we consider laser ablation technique which is one of the most suitable methods for depositing structures of device quality.

As an outstanding example we have to mention the invention relating to a method of exposing a surface on an integrated circuit wafer that involves laser ablation of overlying layers [38]. According to the invention, an integrated circuit is provided having a metal-containing surface feature. A silicon dioxide passivation layer and a silicon nitride passivation layer are provided upon the surface feature and the adjacent integrated circuit surface [38]. A stress buffer containing polyimide is formed upon the silicon nitride passivation layer. Electromagnetic radiation emitted from a laser ablates material successively from the polyimide stress buffer and the silicon nitride passivation layer, thereby exposing an underlying silicon dioxide passivation layer [38]. The silicon dioxide layer is substantially transparent to the radiation emitted from the laser. Consequently, the radiation passes through the silicon dioxide layer and is absorbed by the underlying metal-containing surface feature. Material is ablated from a region of the underlying metal-containing surface feature, which in turn removes material from an overlying region of the silicon dioxide layer. As a result, a surface on the metal-containing surface feature is exposed [38]. The method provides access to various surface features of a semiconductor chip during or after packaging processes. Such surface features include bond pads, fuse lines, and scribe streets. Bond pads processed according to the invention have an exposed surface sufficiently planarized and smooth to adequately bond with external wiring [38].

The method of the invention [38] provides the advantage of exposing a surface on a semiconductor chip while eliminating photo steps. Accordingly, the costly steps of depositing photoresist material, patterning the photoresist material, etching the underlying layers, and stripping the photoresist material are avoided [38]. According to the invention, the polyimide stress buffer has a thickness sufficient to shield the integrated circuit from alpha particle radiation. Using polyimide as an alpha particle barrier eliminates the use of relatively expensive high end molding compound that is conventionally used to absorb alpha particles. The method allows semiconductor chips to function in environments wherein radiation might otherwise reduce reliability, damage memory, or otherwise cause failure [38]. Additionally, the method of this invention provides an option of attaching a lead frame directly upon the polyimide stress buffer [38]. Thus, the need for using lead-on-chip tape is avoided and the cost of post manufacturing packaging processes is reduced. The polyimide conforms to the lead frame such that air pockets trapped against the polyimide are substantially eliminated [38].

The pulse laser deposition (PLD) is successfully applied to fabricating thin films of quaternary solid solutions ZnCdHgTe. The PLD process takes place in a standard vacuum plant equipped with a turbomolecular pump (Fig. (**17**)) [39]. The optical path of the laser beam is shown by dotted lines. Target materials are vaporized by pulsed *YAG:Nd* laser placed in the vicinity of the vacuum chamber [39]. Three different materials can be evaporated from three rotating targets placed side-by-side in the vacuum chamber. The targets are driven from the outside of the vacuum chamber with a help of the magnetic clutch. The angular speed of the targets can be suitably matched to the laser pulse frequency to avoid the overlapping of two sequential laser beam spots on the target surface. The pulse frequency can be chosen between 1 and 50 Hz. Typical working frequency is 10 Hz. The energy of a single pulse can be changed in the range (0.2-.5) J and its duration is 100 ms [39].

The laser beam is focused on the targets by a concave metallic mirror [39]. Between this mirror and the targets, the laser beam is deflected by another (flat) mirror which switches the beam between targets. Both mirrors are covered by a suitable dielectric thin film to increase the reflectivity to near 100% [39].

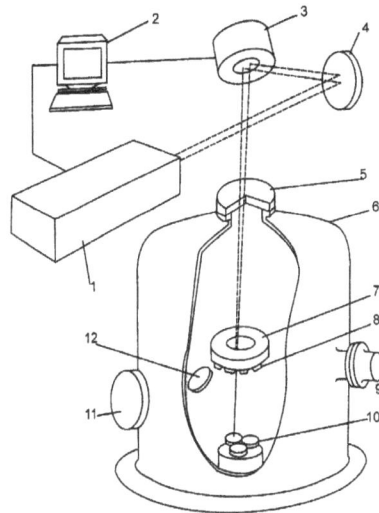

Figure 17: Experimental setup for PLD of the films belonging to the A^2B^6 compounds. Dotted lines show the optical path of the laser beam. (1) *YAG:Nd* laser; (2) computer controlled system of the laser beam monitoring; (3) device switching the laser beam between targets (optical detector); (4) focusing mirror; (5) optical port; (6) vacuum chamber; (7) substrate holder and heater; (8) substrates; (9) quadruple mass spectrometer port; (10) targets setup; (11) peephole; (12) piezoelectric quartz of the film thickness measuring device [39].

The deflecting mirror is placed on a tripod of which each arm can be electromagnetically shortened slightly to deflect the beam in the proper direction. In this way the laser pulses can be switched between the targets. The time needed to switching the beam between targets can be shorter than 100 μs [39].

The evaporated material is condensed on substrates, which are attached to a heater placed above the targets. The targets are pellets, 2 cm in diameter and several millimeters thick, made of powdered material pressed under the pressure of 2.5 MPa. As the laser spot has the diameter of the order of 1 mm, many powder grains are vaporized simultaneously by a single pulse. Under these conditions, a *CdZnTe* target can be prepared from *CdZnTe* powder or from the mixture of *CdTe* and *ZnTe* powders [39].

The process of the laser pulse triggering and the switching between the targets is computer controlled. By appropriate computer program any set of laser pulses deliberately distributed between the three targets can be executed. Such a set can subsequently be repeated many times in sequence. In this way, if the targets are made of different materials, one can obtain a mixed material thin film or a sequence of different films (a superlattice) of compositions corresponding to the chemical compositions of the targets or even a superlattice composed of the mixed material layers [39].

In the experiments described above the laser pulse energy ranged from 0.3 to 0.5 J, and the spot diameter was about 2 mm. As the result it has been found that the available, splashing-free evaporation rate per pulse ranges from 10^{-3} to 10^{-1} nm/pulse [39]. The lower limit corresponds to the evaporation time of several hours for preparation of a film with ~ 0.5 μm thickness [39].

As we have mentioned above, HgCdTe thin films are mostly grown by the liquid phase epitaxy on different substrates: semi-insulating CdTe wafers, CdZnTe solid solution single crystals, GaAs substrates, *etc.* It is inevitable that there exist some defects in $Hg_{1-x}Cd_xTe$ films, such as dislocation, crosshatch, The precipitation [40]. These imperfections can affect the performance of devices based on the material. Nevertheless, the successfully developed modified liquid phase epitaxy and molecular beam epitaxy techniques can provide a quality of the material corresponding to the main advantages of the solid solution HgCdTe [41]: adjustable bandgap from 0.7 to 25 μm; direct bandgap with a high absorption coefficient; moderate dielectric constant/index of refraction; moderate thermal coefficient of expansion; availability of wide bandgap lattice-matched substrates for epitaxial growth.

Further development of the molecular beam epitaxy has allowed preparing nanostructures based on mercury cadmium telluride. Investigation of the fundamental characteristics of HgTe-based quantum wells (QWs) showed [42] that the electron effective mass in HgTe-based QWs is lower than that in QWs of III-V compounds. This leads to a wider splitting of Landau levels, weaker electron localization, increasing the amplitude of Shubnikov–de Haas oscillations, and increasing the Rashba effect [42]. HgTe-based QWs with thicknesses over than 5 nm are characterized by the inverted conduction band (the band of light holes becomes similar to the conduction band) [3]. The typical structure grown by molecular beam epitaxy is shown in Fig. (**18**) [42].

CdTe cap ~ 40 nm	
5 nm undoped	$Cd_{0.7}Hg_{0.3}Te$
10 nm In-doped	
8 nm undoped	
$Cd_xHg_{1-x}Te$ - QW	
8 nm undoped	$Cd_{0.7}Hg_{0.3}Te$
10 nm In-doped	
5 nm undoped	
CdTe/ZnTe/GaAs substrate	

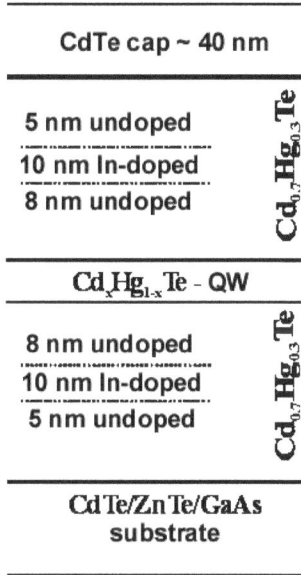

Figure 18: Scheme of a $Cd_xHg_{1-x}Te$-based quantum well [42].

Authors [42] have demonstrated the following results: $Cd_xHg_{1-x}Te$-based (x = 0-0.25) quantum wells (QWs) of 8-22 nm in thickness were grown on (013) CdTe/ZnTe/GaAs substrates by molecular beam epitaxy. The composition and thickness (d) of wide-gap layers (spacers) were $x \sim 0.7$ mol.frac. and $d \sim 35$ nm, respectively, at both sides of the quantum well. The thickness and composition of epilayers during the growth were controlled by ellipsometry *in situ*. It was shown that the accuracy of thickness and composition were $\Delta x = \pm 0.002$, $\Delta d = \pm 0.5$ nm. The central part of spacers (10 nm thick) was doped by indium up to a carrier concentration of $\sim 10^{15}$ cm^{-3}. A CdTe cap layer 40 nm in thickness was grown to protect QW. The compositions of the spacer and QWs were determined by measuring the E_1 and $E_1+\Delta_1$ peaks in reflection spectra using layer-by-layer chemical etching. The galvanomagnetic investigations (the range of magnetic fields was 0-13 T) of the grown QW showed the presence of a 2D electron gas in all the samples. The 2D electron mobility $\mu e = (2.4-3.5) \times 10^5$ cm^2/(V·s) for the concentrations $N = (1.5-3) \times 10^{11}$ cm^{-2} ($x < 0.11$) that confirms a high quality of the grown QWs [42].

In recent years huge advances have been made in HgCdTe focal-plane array (FPA) technology, resulting in the current availability of a wide variety of large-area FPAs, for all IR spectral bands, both monocolor and multicolor, with the capability for both passive and active imaging [43].

The narrow-gap semiconductor solid solution HgCdTe is a material of choice for infrared (IR) optoelectronics. Both bulk and layer materials are served as building blocks of IR-devices. Now we consider important innovations in the field.

Lithographic imaging systems are well known in the prior art for recording images [44]. To do this, a resist coated substrate is provided and then exposed. After exposure, the resist is developed, and then the substrate is etched *via* a chemical and/or physical process to yield the reticulated substrate. For infrared (IR) systems, the substrate that is used is often made from a combination of Mercury, Cadmium and Telluride materials (HgCdTe). After exposure, the resist is developed and an Electron Cyclotron Resonance (ECR) method in an argon/hydrogen gas environment is used to etch the HgCdTe substrate [44].

A lithographic imaging method [44] includes the initial step of providing a substrate made from Mercury, Cadmium and Telluride materials (HgCdTe). The HgCdTe substrate is coated with a diazonaphthoquinone (DNQ) Novolak photoresist material to establish an imaging medium. The imaging medium is exposed to an image pattern and then developed in a tetra-methyl ammonium hydroxide (TMAH) solution. The TMAH solution includes a fullerene (C_{60}) material dissolved therein to retard the subsequent etching of the imaging medium [44]. The incorporation of fullerene into the photoresist material indirectly *via* the developing solution avoids the solubility and ultraviolet (UV) absorbance disadvantages inherent in adding fullerenes directly to the photoresist prior to placement on the substrate. After development, the imaging medium is etched to transfer the recorded image pattern to the substrate. The fullerene cooperates with the photoresist material to slow the etching process, which allows for a highly reticulated HgCdTe detectors and IR images having greatly enhanced resolution [44].

Preparation of device quality substrates and thin films with the lowest number of defects is a technological challenge, especially in case of epitaxial growth of HgCdTe layers designed for IR-detectors. A fruitful way for solving the problem is proposed by authors [45].

Conventionally, as an infrared ray sensor to detect an infrared ray, for example, p-n junction diode in which an epitaxially grown HgCdTe film is formed on CdTe system compound semiconductor substrate such as CdTe substrate, CdZnTe substrate, to be used as a photodiode is known [45]. The substrate for epitaxial growth to form the epitaxially grown HgCdTe film is subjected to a mirror finish [45]. According to [45], the CdTe substrate for epitaxial growth is subjected to the following procedures: CdTe single crystal obtained by a vertical Bridgman method or the like is thinly cut; the thinly cut CdTe single crystal is wrapped, etched, polished, and the like; and eventually final polishing to finish the wafer surface to be a mirror surface is performed [45]. There are for example, mirror etching method and mirror polishing method, as the final polishing which allow obtaining a mirror surface with less waviness and surface anomalies [45].

By the way, the substrate for epitaxial growth which has been subjected to the mirror finish is usually not used for the epitaxial growth immediately as it is, but is used for the epitaxial growth after having been stored for a predetermined period of time [45]. At this point, the substrates for epitaxial growth are required to keep their surfaces clean, and thus, are stored in a state of being housed in a wafer storage container one by one [45].

On the surface of the CdTe substrate which has been stored for the predetermined period of time, not a few oxidized films which prevent normal epitaxial growth are formed. Conventionally, an etching treatment by Br_2 methanol is performed in order to remove the oxidized films as a pretreatment of the epitaxial growth [45].

Authors [45] have proposed a technological process aiming to decrease the generation of defects on the HgCdTe growth film, by providing a semiconductor substrate for epitaxial growth which does not require the etching treatment as the pretreatment, when performing the epitaxial growth of HgCdTe film.

The process includes the following steps [45]: i) the substrate is housed in an inactive gas atmosphere in a predetermined amount of time after the substrate has been subjected to a mirror finish treatment; ii) a proportion of a Te oxide of a total amount of Te on a surface of the substrate is not more than 30%, the proportion being able to be obtained by an XPS measurement. Here, "total amount of Te" means the amount including the simple substance of Te, and Te compound [45].

Thus, by making the growth of the HgTe theamount of Te oxide/total amount of Te on the CdTe substrate surface before epitaxially grown in CdTe film should not exceed 30%, the number of defects generated in the HgCdTe film can be extremely reduced compared to the case in which the etching treatment is performed as the pretreatment [6]. In this case, it is preferable that time until the substrate which has been subjected to the mirror finish is stored in the nitrogen gas atmosphere is not more than 10 hours [45]. Further, the etching treatment need not be performed [45]. Thus, the conventionally performed etching treatment can be omitted [45]. Moreover, the deterioration of the quality of the substrate surface owing to the etching treatment, and the generation of the defects on the surface owing to the deterioration of the surface quality can be prevented. Therefore, the HgCdTe film with extremely fine quality can be grown [45].

The semiconductor $Hg_{1-x}Cd_xTe$ (x=0.3) is proposed as a component of heterojunctions based on ZnCdTe bulk material using for gamma radiation detectors [46].

The resolution for semiconductor-based gamma radiation detectors is defined as the full width at half max of a peak (FWHM) divided by the energy of the peak [46]. The ideal characteristic for this resolution would be an impulse function. This, however, is not typically the case and the signals can be hard to resolve [46]. For semiconductor-based gamma radiation detectors it is the noise within the detector that is responsible for some of the change from the ideal scenario. Currently, high purity germanium (HPGe) detectors offer possibly the best performance for detecting gamma photons and yield a resolution of about 0.2%. However, because of a narrow energy band gap (E_g=0.7 eV), HPGe detectors are operated at cryogenic temperatures to operate properly, typically below 110 K [46]. This low bandgap value allows a relatively large amount of thermally generated current which degrades the signal to noise ratio in the detector, thus prompting the low temperatures of operation. The cooling requirement of Ge is an encumbrance and a room temperature detector would be greatly preferred, to allow for greater portability, operating efficiency, and ease of use [46].

At present, the only commercially available room-temperature (E_g=1.6 eV) alternative to cryogenically-cooled germanium detectors is based on Cadmium Zinc

Telluride (CdZnTe or CZT), which has a resolution of about 10 times greater than Ge based gamma detectors [46]. High resolution gamma detectors may be used for unambiguous identification of special nuclear materials. Fig. (**19**) [46] shows an energy spectrum using three different detector materials, Germanium (Ge), Cadmium Zinc Telluride (CZT), and Sodium Iodide (NaI) exposed to 662 keV gamma energy. Only the Ge material is able to resolve the Special Nuclear Material (SNM) signatures with high enough accuracy, as is evidenced by the narrow and tall peaks [46]. NaI resolution is very poor, as can be seen by the lack of distinct peaks. CZT peaks are evident, but to obtain the resolution desired for use in discovering unambiguously special nuclear materials, the peaks should be taller and the width of the peak should be narrower [46]. This sort of peak probably cannot be realized with CZT unless the dark current in the sensor is minimized somehow. Dark current, or leakage current, in a gamma detector is the amount of electrons and/or holes that enter the semiconductor material used to detect gamma radiation from electrodes thereof and travel across the semiconductor material or the amount of electrons and holes thermally generated in the bulk material. Dark current decreases the performance of any gamma detector material. However, improvement in the CdZnTe may result in a detector material that can resolve the SNM signatures in a similar capacity as that of Ge [46].

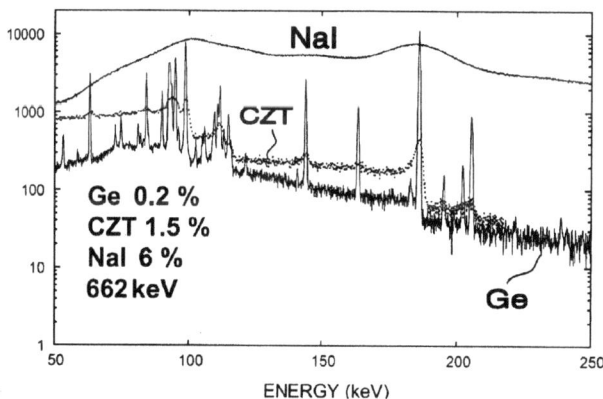

Figure 19: A plot of energy resolution for three materials: NaI, CZT (CdZnTe), and Ge [46].

Therefore, a CdZnTe-based gamma radiation detector that can operate effectively at or near room temperatures and still provide suitable resolution would be very beneficial [46].

A method for choosing a barrier material for a gamma detector comprises determining a lattice constant for a semiconductor gamma detector material and a first material to be used as a barrier material, determining if the lattice constant of the semiconductor gamma detector material and the lattice constant of the first material are within about 10% of each other, wherein if the lattice constants of the first material and the semiconductor gamma detector material are within about 10% of each other: determining an energy barrier to electron movement across the first material and across the semiconductor gamma detector material, determining an energy barrier to hole movement across the first material and across the semiconductor gamma detector material, determining if the energy barrier to electron movement across the first material is higher than the energy barrier to electron movement across the semiconductor gamma detector material, wherein if the energy barrier to electron movement across the first material is higher than the energy barrier to electron movement across the semiconductor gamma detector material, the first material is selected as an electron barrier material to be used with the semiconductor gamma detector material in a gamma detector, determining if the energy barrier to hole movement across the first material is higher than the energy barrier to hole movement across the semiconductor gamma detector material, wherein if the energy barrier to hole movement across the first material is higher than the energy barrier to hole movement across the semiconductor gamma detector material, the first material is selected as a hole barrier material to be used with the semiconductor gamma detector material in a gamma detector [46].

There are two approaches that can be taken to improve the resolution of the CdZnTe-based gamma detectors: first, the resistivity of the CdZnTe material can be increased which may reduce the current through the device, and second, the physical layer design of the device may be changed. In one approach, latticed-matched materials that may be grown on CdZnTe detectors may be used to block the leakage current within the device. These lattice-matched materials may be used in a gamma detector system to create high resolution CdZnTe-based gamma detectors [46].

There are several components contributed to the "noise" which can hinder the resolution of CdZnTe gamma detectors and gamma detectors using other materials as a gamma detector material, including material non-uniformity, recombination of carriers from traps, read out electronics, incomplete charge carrier collection, and/or the leakage current through the device [46]. The dominant component of noise for conventional detectors with ohmic contacts is the leakage current through the device from these contacts, and thus this source of noise is reduced according to embodiments disclosed herein, since this can have some of the greatest impact on improving the performance of gamma radiation detectors [46].

To overcome the limitations of an ohmic detector structure, effective reverse biased Schottky barrier diodes may be used to block the leakage current, according to some embodiments [46]. Thus, considerably higher electric fields can be applied to increase collection of the created carriers due to the reduced noise. The simplistic thermionic emission theory calculates the reverse bias current density as being dependent on the barrier height between the semiconductor and the metal, according to Eq. (**2.1**) [46]:

$$J_r = -A^* T^2 \exp(-q\Phi_b/k_B T) \tag{2.1}$$

where J_r is the emission current density, A^* is the Richardson constant, T is the temperature of the metal, k_B is the Boltzmann constant, q is the charge on an electron, and Φ_b is the difference between the work function of the metal and the electron affinity of the semiconductor [7]. However, nature has provided a more complex situation where the barrier height may be reduced due to interface effects between the semiconductor and the metal, which include Fermi level pinning and surface leakage current [46].

To achieve a goal of less than about 1% resolution, the leakage current should be below about 3 nA/cm^2 [46]. According to some novel approaches, in order to accomplish this goal, a multilayered contact structure may be used and intermediate materials may be inserted between the semiconductor and the metal to suppress the surface effects and increase the Schottky barrier height [46]. According to one approach, a lattice-matched semiconductor epitaxial growth may be used on top of CdZnTe crystals to form heterojunctions to substantially block the leakage current but not the signal carrying charged carriers, such as electrons (negatively charged) and holes (positively charged). Epitaxial growth indicates that the deposited film or upper layer takes on the lattice structure of the substrate or lower layer [46].

The leakage current due to electrons originates from the cathode side of the detector but by incorporating HgCdTe, the electrons may be blocked by a 1 eV energy barrier that forms at the heterojunction between CdZnTe and HgCdTe [46]. An analogous situation can be engineered at the anode side of the structure by using InSb on CdZnTe to block the holes. The relationship between leakage current and energy barriers is shown in Fig. (**20**) [46] which shows that for a 70% Hg composition, the leakage current is reduced by 10^8 orders of magnitude assuming thermionic emission over the barrier [46].

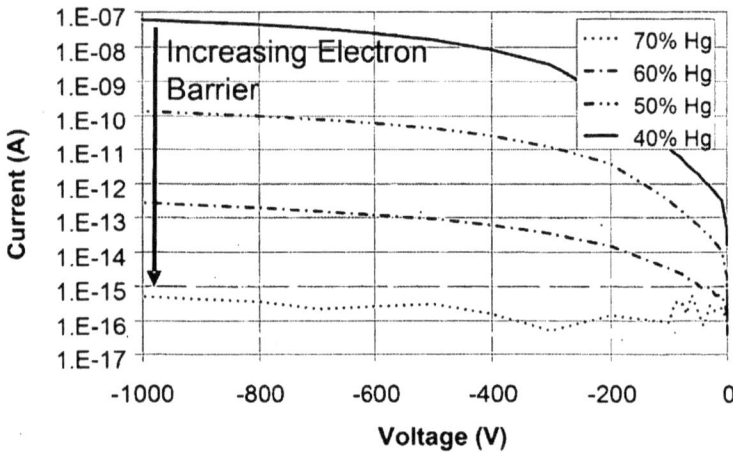

Figure 20: Reverse current-voltage characteristic of the device with different content of Hg in the HgCdTe component [46].

An added advantage of Schottky contacts may be the application of a higher applied field (E). Since the electron range is given as $\mu\tau E$, where μ is the mobility and τ is the carrier lifetime, a higher applied field (E) should be able to improve the carrier collection efficiency and therefore the resolution [46]. Conventional epitaxial crystal growth methods such as molecular beam epitaxy (MBE) and metalorganic chemical vapor deposition (MOCVD) can be used to fabricate the latticed-matched structures [46]. Epitaxial growth indicates that the deposited film or upper layer takes on the lattice structure of the substrate or lower layer [46].

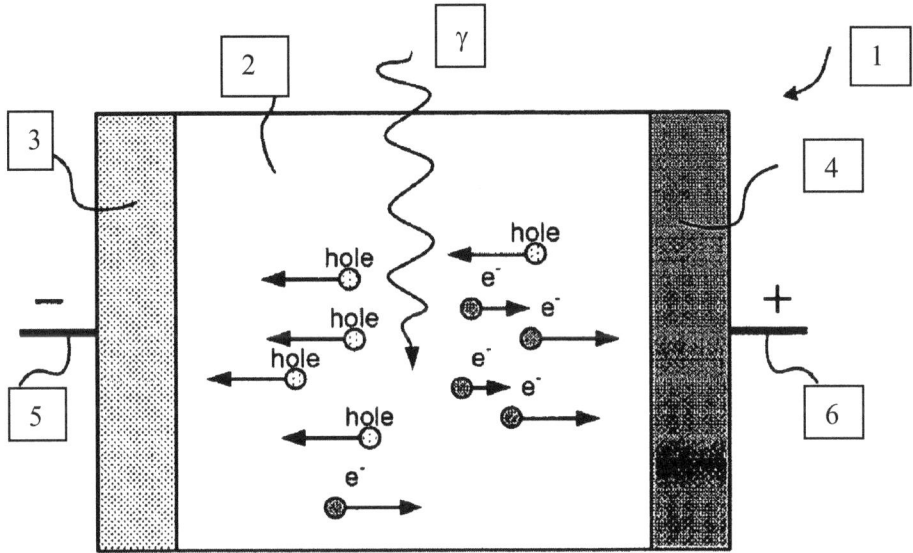

Figure 21: Schematic diagram (1) of a semiconductor gamma radiation detector [7]: (2)-semiconductor γdetector material, (3)-electron blocking layer, (4)-hole blocking layer, (5) and (6)-first and second electrodes of the external voltage source [46].

The schematic diagram of the device is shown in Fig. (**21**) [46]. The system 1 includes a semiconductor gamma detector material 2, a first electrode 5 adjacent the gamma detector material 2, and an electron blocking layer 3 positioned between the gamma detector material 2 and the first electrode 5 such that the first electrode 5 is not in electrical communication with the gamma detector material 2 except *via* the electron blocking layer 3 [46]. The electron blocking layer 3 resists passage of electrons there through [46]. The system 1 also includes a second electrode 6 adjacent (either directly or indirectly adjacent) the gamma detector material 2, and a hole blocking layer 4 positioned between the gamma detector material 2 and the second electrode 6 such that the second electrode 6 is not in electrical communication with the gamma detector material 2 except *via* the hole blocking layer 4. The hole blocking layer 4 resists passage of holes there through, and the blocking layers (3 and 4) are lattice-matched to the gamma detector material 2 [46].

An electrode, such as first electrode 5 and/or second electrode 6, may be a contact point on the structure, *e.g.*, the electrode may be a surface of the blocking layer itself, the electrode may be an electrically conductive pad and/or a layer coupled to the nearest blocking layer (*e.g.*, hole blocking layer 4), *etc.* [46].

The semiconductor gamma detector material 2 may comprise CdTe, CdZnTe, and/or combinations of these materials, with component elements in any ratio, as well as the hole blocking layer 4 may comprise InSb, in any ratio of In to Sb, and/or the electron blocking layer 3 may comprise HgCdTe, in any ratio of Hg to Cd to Te, such as $Hg_{0.7}Cd_{0.3}Te$ [46].

The electron blocking layer 3 may provide an equivalent Schottky barrier to electrons of greater than about 0.8 eV, and the hole blocking layer 4 may provide an equivalent Schottky barrier to holes of greater than about 0.8 eV. Further, a resistivity of the gamma detector material 2 may be less than about 10^{11} ohm·cm, more preferably less than about 10^{10} ohm·cm, and even more preferably less than about 10^{9} ohm·cm [46].

Before going to the technology of IR-detectors based on HgCdTe semiconductors we discuss some questions concerning the external electric field-induced characteristics and environmental effects on performance of the structures prepared by methods presented above.

6. HEAT TRANSFER MODELING IN METAL-SEMICONDUCTOR STRUCTURES BASED ON NARROW-GAP COMPOUNDS

Different semiconductor structures (diodes, photodetectors, optoelectronic active elements, metal-semiconductor structures) based on various materials, in particular, A^2B^6 compounds (ZnTe, CdTe, *etc.*) play an outstanding role in novel microelectronics [47]. Their operation under electric fields applied to the active element is ensured by metallic (or intermetallic) contacts formed by means of varied techniques (vacuum deposition and fusion are of special interest due to simplicity and reliability of the methods). Nevertheless, even the best contacts provide a sufficient heat flow through the operation area of the semiconductor heterostructure which can lead to the significant changes of the some principal characteristics of the device, especially, the non-ideality of the current-voltage dependence determining the main processes of the charge transport [47].

Here we discuss results of the studies performed for heating processes in the structure In-ZnTe/In-$Zn_{0.86}Hg_{0.14}$Te.

Samples chosen for experimental measurements were cut off from the as-grown ring plates as rectangular boxes of electric area ~4.5 mm^2. Indium contacts were deposited as homogeneous pads on both left and right end surfaces of the sample.

Current-voltage (I-V) examinations performed under the room temperature were demonstrated the exponential character of the I-V-dependence (Fig. (**21**)) describing by the equation (3.1):

$$I = I_s\left(\exp\left(\frac{eV}{nk_BT}\right) - 1\right),$$

(3.1)

where I_s is a saturation current (a complete function determined by the electronic parameters of the components of the structure), V is an applied voltage, e is an elemental charge, T is the experiment temperature, k_B is the Boltzmann constant and n is a non-ideality factor strongly depended on the effects of the structure warming, and a presence of the quasi-neutral region localized at the ZnTe-ZnHgTe interface [47]:

$$n = \frac{e}{k_BT}\frac{dV}{d\ln I}.$$

(3.2)

Numerical analysis of the experimental data carried out according to the methods developed in [47] pointed out significant non-ideality of the I-V characteristic caused by different heat effects arisen due to charge carriers warming [47]. Fig. (**22**) plots the scheme of the investigated structure using in the following presentation.

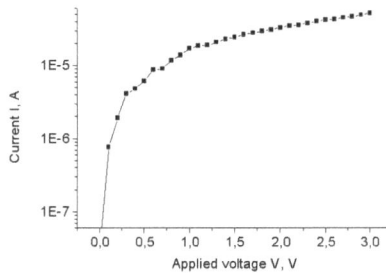

Figure 22: Experimental current-voltage characteristic of the investigated structure, $T_{thermostat}$ = 290 K [47].

The numerical value of *n* in our case changes from n=1 under applied electric field (0-5·10^2) V m^{-1} to n=3 as the electric field increases (Fig. (**21**)). As it is known [47], such values of the non-ideality coefficient indicate the considerable changes in the mechanism of the charge carriers transport in the investigated structure from diffusion one to the tunneling through

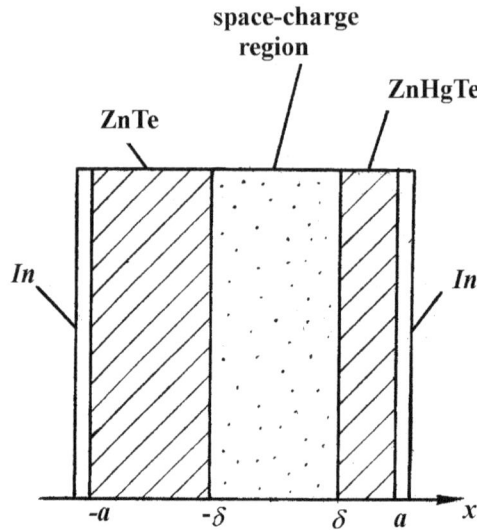

Figure 23: Schematic plot of the investigated active elment: two metal-semiconductor structures In/ZnTe and In/ZnHgTe, as well as the space-charge region [47].

deep charge centers localized at the interface *substrate-film*. These processes are caused by the fluctuations of the charge carrier concentration [47] due to thermal generation from the metal-semiconductor structure on both sides of the sample (see Fig. (**21**)). So, taking into account processes of the heating of the two subsystems (electrons and photons) [47] one can perform the numerical simulation of the experimental results according to Eq. (3.3) in assumption that the current (heating) contacts are placed at *x=±a* under the same temperature (temperature of the thermostat):

$$\frac{eV}{nk_BT} = \ln\left\{\frac{j_s}{j_{s0}}\left[\exp\left(\frac{e\varphi_b}{k_BT_{ph}} - \frac{e(\varphi_b - V)}{k_BT_e}\right) - 1\right] + 1\right\}, \tag{3.3}$$

where j_{s0} is the saturation current without field, φ_b is the potential barrier height (determined experimentally by the capacitance-voltage and current-voltage studies), T_e and T_{ph} are temperatures of electron and phonon subsystems, respectively. The theoretical I-V dependencies are described by the following equations:

$$I_1^{theor} = D_{tun} A_{el} j_{so} \exp\left(\frac{eV}{n_1 k_B T_{th}}\right),$$ **(3.4)**

$$I_2^{theor} = D_{tun} A_{el} j_s \exp\left(\frac{eV}{n_2 k_B T_{th}}\right),$$ **(3.5)**

where D_{tun} stands for the tunneling transparency of the potential barrier at the interface ZnTe/ZnHgTe and is modulated by the thermally generated from the metal-semiconductor regions carriers, A_{el} is an electrical area of the sample (the area covered by the heat and charge carriers flows), T_{th} takes care of the thermostat temperature (determined by the conditions of the experiment). The case when temperatures of the electron and photon subsystems are equal to that of the thermostat under applied electric fields no more than $5 \cdot 10^2$ V m^{-1} is seemed to be of special interest for practical requirements. So, results of the numerical modeling are plotted in Fig. (**23**) [47].

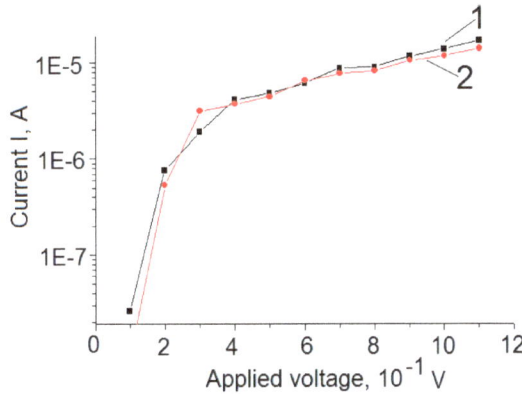

Figure 24: Theoretical I-V-dependence (curve 2) calculated according to the Eqs. (4), (5) in comparison with the experimental data (curve 1) [47].

The data obtained from the experimental examinations and numerical modeling of electric field-induced heat effects in the metal-semiconductor structure based on wide-gap A^2B^6 materials demonstrate the sufficient influence of the thermally generated carriers on the current-voltage characteristic and the height of the potential barrier of the structure, in particular, on the value of the tunneling transparency coefficient D_{tun} which varies from 10^{-8} to 10^{-4}. This fact can be explained by the thermal flip-over of the carriers' packets under the definite values of the applied electric field and appearance of the peculiar waves of the fluctuation of the charge carrier concentration [47].

7. NUMERICAL SIMULATION OF PARAMETERS OF ZnCdHgTe FILMS AND ZnCdHgTe-BASED HETEROSTRUCTURES

The epitaxial ZnCdHgTe films of 8-15 µm thickness and electrical area 2-5 mm^2 grown by the modified liquid-phase epitaxy on the (111)-oriented monocrystalline CdTe wafers were selected for electric-field measurements performed under the room temperature in the range of applied bias (0-200) mV.

The main parameters of the grown epilayers and heterostructures ZnCdHgTe/CdTe are listed in Table **1** [48].

Table 1: Some parameters used in calculations

Parameter	ZnCdHgTe	CdTe
Dielectric constant	16.3 ε_0	10.9 ε_0
Electron effective mass	0.01 m_0	0.11 m_0
Hole effective mass	0.55 m_0	0.35 m_0
Lattice parameter a, nm	0.6392	0.6478
Band-gap E_g, eV	0.32	1.44

Peculiarities of the experimental current-voltage characteristics measured under the conditions described above were shown a dominant tunneling current flowing through the the epilayer surface. Numerical simulation performed according to the experimental data has demonstrated the significant influence of the randomized relief of the epilayer [48-50]:

$$I = \left(\frac{R}{r}\right) I_s \exp\left(-\alpha\varphi(r)\right)\exp\left[-e(E_g - V_a)/k_B T\right], \tag{4.1}$$

where $\alpha = (4/3\hbar)(m_e^* \varepsilon\varepsilon_0 / \delta N)^{1/2}$, r is a variable determined by the clusters size range, I_s stands for the saturation current (a complete function strongly depending on the main parameters of the material), E_g is the gap of ZnCdHgTe film (this value is a composition function), R is the cluster size, δN is the concentration of carriers immediately taking part in process of the charge transport evaluated by the experimental capacitance-voltage measurements, V_a denotes the applied bias. The randomized potential $\phi(r)$ was estimated according to the expression given below:

$$\varphi(r) = a(r_0^2)\left[1 - (r_0/r)(1 + R/r_0)\exp(-R/r_0)sh(r/r_0)\right], \tag{4.2}$$

where $a = e\delta N/\varepsilon\varepsilon_0$, $(r_0)^{-2} = n_i e^2/\varepsilon\varepsilon_0 k_B T$; both parameters are determined by the properties of ZnCdHgTe epifilm, n_i stands for intrinsic carriers concentration calculated in assumption of parabolicity of the band structure of the solid solution ZnCdHgTe [48, 49].

Two theoretical models (percolation theory and the model of so-called "p-n"-island conductivity [48]) were used for numerical modelling.

Sections of theoretical and experimental curves (Fig. (**24**)) [48] are well matched up to applied voltage (0-0.15) V. The percolation theory shows the dramatic increase of the calculated values of the current as the applied voltage increases. Such a phenomenon can be explained by formation of the conductive channels localized along the clusters boundaries. Nevertheless, the "p-n-island" conductivity theory gives no similar results due to the protection influence of the native oxide film coating the as-grown epilayers immediately after their growth finishing.

A simple numerical technique has worked out for calculations of the main parameters of multi-component semiconductor heterostructures [49]. In particular, the procedure deals with numerical treatment of room-temperature capacitance-voltage measurements having for an object the determination of some properties of the semiconductor free surface. Thus, this part of the paper reports first results of calculation performed for determination not only the electron affinity but the work function of the sufficiently inhomogeneous surface of ZnCdHgTe epitaxial layer.

The proposed numerical model was refined by means of electron affinity calculation and consequent work function definition because these parameters are seemed to be important for construction of the band diagram of the tested semiconductor heterostructure as a base for further design of microelectronics active elements.

Capacitance-voltage ($C=f(V_a)$) measurements were carried out at T = 290 K. Results of the study (Fig. (**25**)) allowed to estimate the energy spectrum of charge surface states localized at the interface *ZnCdHgTe-CdTe-substrate* (E_{ss},) and to define a unique electrostatic potential (V_c). Thus, the experiment gives sufficient information to solve numerically the basic equation (4.1) [50]:

$$V_c = V_a - \varphi_{bp} - \chi_s + \chi_{layer} - E_{gs} + E_{ss} + \frac{k_B T}{2e} \ln\left(\frac{N_{clayer}}{N_{vlayer}} \exp\left(-\frac{E_{cd}}{k_B T} \right) \right), \qquad (4.3)$$

where ϕ_{bp} is the flat-band voltage, χ_s and χ_{layer} are the substrate and ZnCdhgTe epilayer electron affinities, respectively, E_{gs} is band-gap of the substrate, N_c and N_v are the effective states densities in the conductance and valence bands of the epilayer. The computing scheme is presented below:

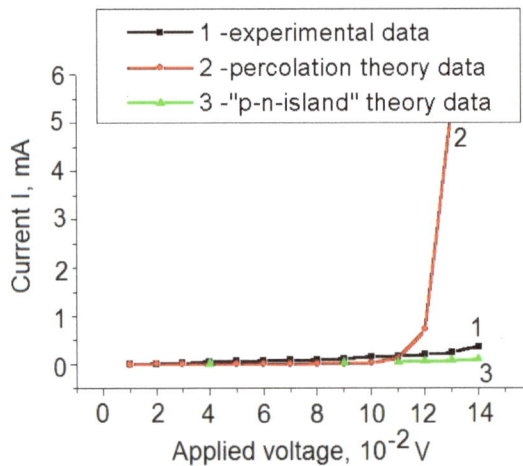

Figure 25: Experimental (curve 1) and simulated (curves 2 and 3, respectively) current-voltage characteristics of the investigated material. As numerical simulation tools the percolation theory and "p-n-island" conductivity models were used. The second theory produces more reliable data giving no sharp increase of the current [48].

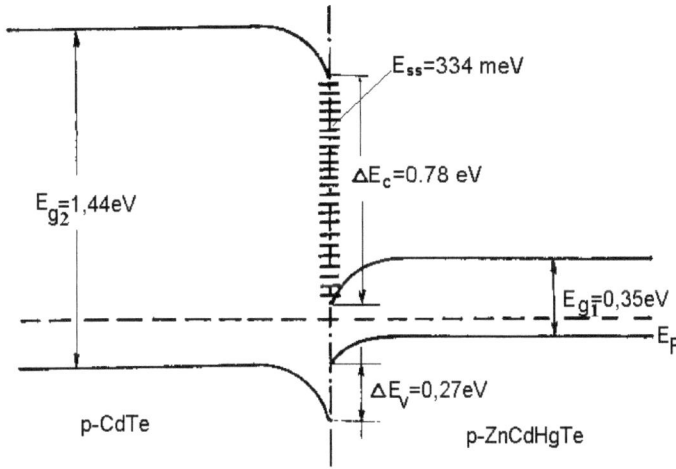

Figure 26: Energy band diagram of the *p-p*-heterostructure based on the narrow-gap semiconductor ZnCdHgTe and flowchart of the calculation algorithm (below) [48].

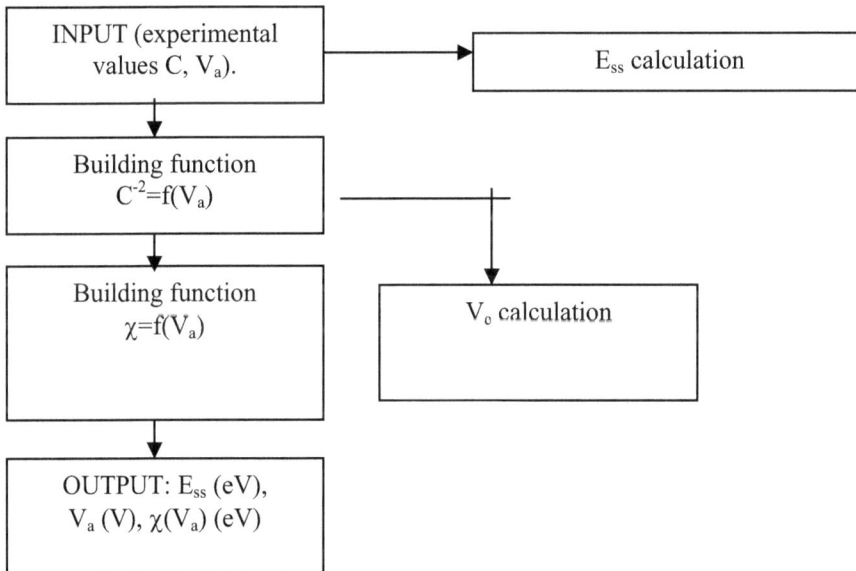

Thus, the described numerical simulation can give a possibility to determine the following factors: (i) some important technological parameters of the modified LPE technique in order to avoid undesirable surface relief of the as-grown film; (ii) results of the calculations make it possible to determine the space distribution distribution of the charge carriers and the film surface mobility modulation [48].

8. PECULIARITIES OF CURRENT-VOLTAGE CHARACTERISTICS OF EPITAXIAL ZnCdHgTe FILMS DUE TO INFLUENCE OF THE ATMOSPHERIC OXYGEN

Taking into account that the semiconductor quaternary solid solution ZnCdHgTe is a prospective material for thin film-based IR photodetectors, the long-term stability of electric field-induced characteristics of these films after 3 years storage at normal atmospheric conditions is seemed to be of special interest.

The best samples of ZnCdHgTe epilayers grown on CdTe:Zn substrate of (111) orientation with the film thickness of ~10-12μm, the free carrier concentration 7×10^{14}-5×10^{15}cm^{-3} and mobility $(0.4-5) \times 10^{3}$cm^{2}V^{-1}s^{-1} at the temperature range 77-200K were selected for electrical investigations. Electrical area of the samples was estimated to be about (3-5) $\times 10^{-2}$cm^{2}. Various material characterization techniques such as transmission and scanning electron microscopy, IR transmission spectroscopy were used o study the structural properties, kinetics of formation, morphology and composition of epitaxial ZnCdHgTe epilayers [51]. We have found tessellated structure on the surface of these films, without composition changes which could be related with the so-called time degradation of this solid solution [51].

Fig. (27) shows room-temperature current-voltage characteristics (IVC) for as-grown p-Zn$_{0.12}$Cd$_{0.18}$Hg$_{0.70}$Te epilayer as well as the same characteristics measured after 3 years storage of the samples at the normal atmospheric conditions [51]. It is obvious that the IVC's evolution indicates some interesting facts: (i) the appearance of exponential dependence $j \sim exp(eV_{a}/nk_{B}T)$, where the ideality factor n is ranged from 2.31 to 8.83, and (ii) arising new carrier transport mechanism [51]. To explain the results observed the theoretical models involving tunneling channels formed at the interface in the direction *film surface-film depth* and the appearance of the space charge at the narrow subsurface region of the epitaxial layer due to fluctuation of the charge centers' concentration were used [51].

The experimental results demonstrated the pronounced change of the current mechanism: instead of the carriers' tunneling (indicating by the significantly high values of the ideality factor *n*) the hopping conductivity mechanism starts to dominate [51]. Table **2** [51] lists some parameters calculated from the experimental IVCs.

Table 2: Numerical parameters extracted from the experimental current-voltage characteristics of the p-ZnCdHgTe/p-CdTe heterostructure [51]

Bohr radius a_B, m	Bohr energy E_B, meV	Tunneling channel length x_t, μm	E_0, meV	I_h/I_t
1.52×10^{-9}	30.4	9.6	0.005	1.56×10^{-2}

In particular, it has been found that the tunneling channels were formed almost in the depth of the ZnCdHgTe film [52-56]:

$$x_t = (V_d)^{(7/6)-m}(E_0)^{m/3}a_B/[N_A(a_B)^3]^{2/3}(E_B)^{7/6},\qquad(5.1)$$

where V_d is the barrier height determined from the experimental IVC, a_B and E_B are the Bohr radius and the Bohr energy, respectively, E_0 is the energy level of the charge fluctuations, and *m* is a numerical parameter ($m \sim 0.85$). On the other hand, the charged defects activated by the atmospheric oxygen are appearing as those with sufficient concentration [52-56]:

$$N_0 = N_A\{[N_A(a_B)^3]^{1/3}(E_B)^{4/3}\}/[(V_d)^{(1-2m)/3}](E_0)^{2m/3},\qquad(5.2)$$

where N_A is the acceptor concentration in the ZnCdHgTe film [52-56].

We should note that the capacitance-voltage measurements have resulted in the similar values of x_t and N_0: $x_t \sim 0.18$μm, and $N_0 \sim 1.54\times10^{13}$cm^{-3} [51]. To verify the suggestion about the current mechanism caused by the charge carriers' hopping conductivity we have calculated the ratio of the tunneling current I_t and the hopping current I_h [52-56]:

$$I_h/I_t = \{15\pi^{1/6}[N_A(a_B)^3]^{1/6}/2^{5/6}(5+2m)^{2/3}\} \times (E_B/V_d)^{(5-4m)1/2}(E_0/E_B)^{m/3}\qquad(5.3)$$

It has been found that $(I_h/I_t) = 1.56\times10^{-2} \ll 1$. Thus, we have observed the hopping conductance current mechanism [51]. The following analysis of the experimental IVCs enabled us to propose an energy band diagram of the elemental heterostructure p-Zn$_{0.12}$Cd$_{0.18}$Hg$_{0.70}$Te/p-CdTe (Fig. (**27**) [51]). The category *elemental heterostructure* here means that the system of such heterojunctions has been appeared and the intercrystalline boundaries in the bulk of ZnCdHgTe film are insulators [51-56].

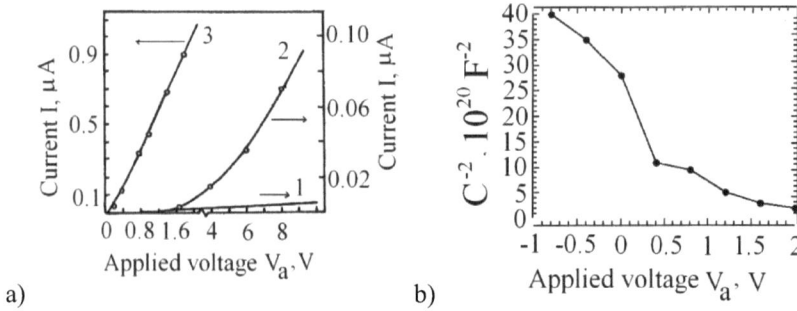

Figure 27: a) Current-voltage characteristic (IVC) of p-ZnCdHgTe epitaxial film: (1)-as-grown layer, (2)-after 1 year of O_2 exposition, (3)-after 3 years storage at normal atmospheric conditions; (b) capacitance-voltage characteristic (CVC) of the investigated sample after 3 years storage at normal atmospheric conditions and under the test signal frequency f = 1kHz [51].

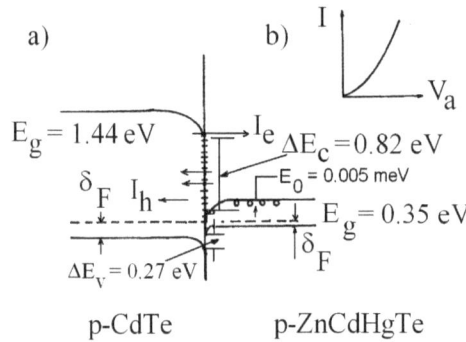

Figure 28: Energy band diagram of the elemental heterojunction p-CdTe/p-ZnCdHgTe with the current directions (a) and the theoretical forward IVC (inset b), T = 290 K [51].

Table 3: Parameters used for the calculation of the energy band diagram of the heterostructure p-ZnCdHgTe/p-CdTe [51]

	Lattice constant a (nm)	Charge carrier concentration (cm^{-3})	E_g (eV)	m_h	ε	χ (eV)
1	0.65	5.5×10^{13}	1.44	$0.35m_0$	10.9	4.48
2	0.64	5.0×10^{16}	0.35	$0.55m_0$	16.0	5.10

Notes: 1-CdTe, 2-ZnCdHgTe, m_h is the hole effective mass (in units of the electron mass m_0) in both components of the heterostructure, χ is the electron affinity of the components of the structure.

The features of the capacitance-voltage characteristic (CVC) (Fig. **(26b)**) studied after 3 years of the sample's exposition to the atmospheric oxygen gave a possibility to estimate the charge and concentration of the electron states localized at the interface of the investigated structure. Assuming that the effective value of the band discontinuity $\Delta\Phi$ does not coincide with the real conduction band discontinuity ΔE_c at the hetero-interface, we obtain [51-56]:

$$\Delta\Phi = \Delta E_c + (k_B T/e)\ln(N_1 N_{c2}/N_2 N_{c1}), \tag{5.4}$$

where N_1 and N_2 are concentrations of the charge carriers in CdTe and ZnCdHgTe, respectively, N_{c1} and N_{c2} are the effective state densities in the conduction bands of CdTe and ZnCdHgTe, $\Delta E_c = |\chi_{ZnCdHgTe} - \chi_{CdTe}| = 0.82$ eV according to the classical Anderson theory for abrupt heterostructures [52-55]. Then from the equation (8.5) [52-56]:

$$(4\pi e N_s)/\varepsilon_1 = V_2 - (2\pi e N_1(x_1)^2)/\varepsilon_1 - V_d - \Delta\Phi, \tag{5.5}$$

where V_2 is a so-called "plateau" voltage and ε_1 is the CdTe dielectric constant, we obtain the electron surface states density $N_{ss} = 1.92 \times 10^{14}$ cm^{-2}.

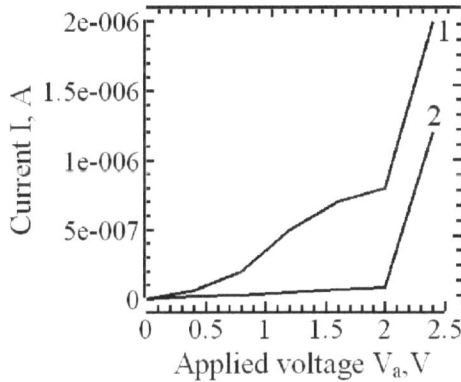

Figure 29: Experimental (curve 1) and theoretical (curve 2) IVCs of the investigated heterostruture [51].

Fig. **(29)** illustrates the results of numerical simulation based on the experimental data (curve 3, Fig. **(28a)** and Table **3**). For the lower applied voltages $0 < V_a < 0.4$ V we have the following expression taking into account the Auger-ionization of the holes in the conduction band of the wide-gap substrate [52-56]:

$$j=PA^*T^2(eV_d/k_BT)(L_{10}/2h)(1+h^2/(L_1)^2)x(\pi\varphi_{00}th\theta_m/\varphi_0ch^2\theta_m)^{1/2}exp[(-e\varphi_s/k_BT)-(eV_d/k_BT)]x$$

$$[1-.083(e\varphi_{00}/k_BT)^2(1+h^2(L_1)^2)(L_1)^2/h^2]exp(eV_a/nk_BT) \qquad (5.6)$$

where $(L_{10})^2 = 2\varepsilon\varepsilon_0\varphi_0/eN_0$, $(L_1)^2 = 2\varepsilon\varepsilon_0V_d/eN_0$, $\varphi_{00} = (h/2)(N_0/m^*\varepsilon\varepsilon_0)^{1/2}$, φ_0 is a work function of the ZnCdHgTe compound, $\theta_m = (e\varphi_{00}/k_BT) << 0.3$ is a numerical parameter, φ_s is the difference between the Fermi level and the bottom of the conduction band [51-56], P is a probability of the Auger ionization of the holes in CdTe (P ~ 10^{-4}) [51]. For $V_a > 0.45$ V the experimental IVC was numerically simulated by the following expression [52-56]:

$$j = en\mu V_a(L)^{-1}exp(\varepsilon_2eV_dd/k_BTL)/[1+(4\mu V_a/v_{th}LP_1)] \qquad (5.7)$$

where L is the electrical length of the sample, ε_2 is the dielectric constant of ZnCdHgTe, v_{th} is the thermal velocity of the charge carriers in the epitaxial film ZnCdHgTe, and the parameters *n, d, P₁* are determined as follows [52-56]:

$$n=j_{min}Lexp\{(1-(\alpha_{xmin})^{0.5})[(1-(\alpha_{xmin})^{0.5})-1]\}/e\mu V_{amin}[(1-(\alpha_{xmin})^{0.5})-1] \qquad (5.8a)$$

$$d=(Lk_BT/e\varepsilon_2V_{amin})(1-(\alpha_{xmin})^{0.5}[1-(1-\alpha_{xmin})^{0.5}] \qquad (5.8b)$$

$$P_1=(4V_{amin}\mu/v_{th}L)[(1-(1-\alpha_{xmin})^{0.5})/(1-\alpha_{xmin})^{0.5}] \qquad (5.8c)$$

$$\alpha_{xmin} = (V_a/j_{min})(dj/dV_a) \qquad (5.8d)$$

As is obvious, there is a discrepancy between experimental and theoretical IVCs (Fig. (28)). It can be explained by the high ideality factor *n* and a low barrier height indicating the presence of defects in a thin layer near the CdTe surface which lowers the values of V_d (especially under the applied bias larger than 1.0 V). These defects can also be a source of an additional leakage current [51].

9. INFRARED PHOTODETECTORS BASED ON NARROW GAP SEMICONDUCTORS

Quantum photoelement function is based on the intrinsic photoeffect in semiconductor. Under this effect the radiation quanta generate charge carriers bounded with the atoms of the crystal lattice or with atoms of impurities. The intrinsic or impurity conductivity arises. In the first case the energy of the radiation quantum should be no less than the energy gap of the semiconductor at the given temperature ($hv{\geq}E_g$). In case of the impurity conductivty the quantum of radiation hast o pessess an energy no less than that of the impurity activation energy ($hv{\geq}E_a$). The known expression gives the relationship oft the quantum energy (in eV) and the radiation wavelength (in μm):

$$\lambda = \frac{1.24}{E(eV)}(\mu m).$$ **(6.1)**

Obvious, that for an intrinsic photodetector with the gap E_g = *0.124 eV* the longwave limit of the photosensitivity is λ_c = *10 μm*. For the impurity photodetector the activation energy of the impurity is to be in this range. However, the impurity photodetectors require deep low-temperature cooling (down to 4.2 K) in opposite to intrinsic photoelements working at the temperature of liquid nitrogen.

Quantum photoelements use the following kinds of the intrinsic photoefffect:

1. Photoconductivity: the conductivity of the semiconductor increases under effective absorbtion of the radiation. There is a principal working function of photoresistors (PR).

2. Photovoltaic effect: it appears in the semiconductor with a potential barrier (*p-n*-junction, heterojunction). In this case the charge carriers generated by incident light are separated by the field of the built-in potential barrier giving rise an electromotive force at the contacts of the device. These active elements can operate without external bias source (photpgalvanic mode) as well as photodiodes if the external source biased in reverse direction is connected with the structure.

3. The photoelectromagnetic effect: the electromotive force appears when the diffusion flow of the charge carriers generated by the radiation moves into depth of the semiconductor under transverse magnetic field spatially separating the charge carriers.

Now we discuss how to estimate the parameters of photodetectors.

Photonic radiation detecors (Fig. (**30**)) are the most usable in the infrared optoelectronics. They are the selective photodetectors based on semiconductor materials with various forbidden gaps. One of the main properties of the photodetector is its spectral characteristic. This function defines how the realative sensitivity of the device depends on the radiation wavelength (Fig. (**30**)).

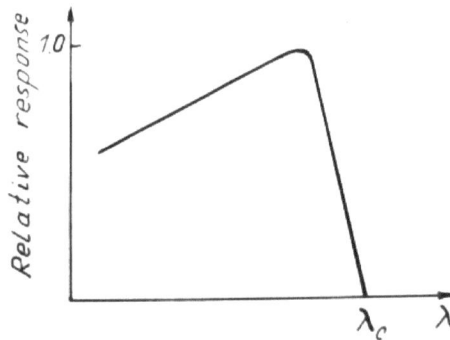

Figure 30: Qualitative spectral characteristic of the photonic photodetector (λ_c is a longwave edge of the photosensitivity).

Almost all photodetectors register radiation passing through the earth atmosphere. The thermal radiation is weakened due to absorption and scattering by the molecules of atmospheric air. The most important atmospheric windows are 2-2.5 µm, 3-5.5 µm and 8-14 µm. The uncooled photodetectors successfully operate in the wavelength ranges corresponding to first two windows, the cooled intrinsic photodetectors based on solid solutions $Cd_xHg_{1-x}Te$ and $Sn_xPb_{1-x}Te$ function in the spectral interval 8-14 µm. CO_2-lasers also work in this wavelength interval ($\lambda = 10.6$ µm).

Every photodevice is a structure with input (incident electromagnetic radiation) and output (electric signal V_s as a result of incident radiation transformation). In real devices the spontaneous fluctuations (noises) of the output signals are of particular importance.

The detectivity of the device is defined as a ratio of the output signal to the power of the incident radiation:

$R = V_s/P$ $\qquad\qquad$ **(6.2)**

In this expression V_s is an averaged square of the min signal (without harmonics)-P is a power of the incident radiation; this value is a product of the radiation flow density $H(J/sm^2)$ and the device area A: $P = HA$.

The next parameter is a density of the incident radiation flow forming at the detector output a signal equivalent to the signal of a noise normalized to the unit interval Δf:

$$H_n = \frac{V_n}{A(\Delta f)^{1/2} R}$$ $\qquad\qquad$ **(6.3)**

where V_n is an averaged square of the noise bias. The product of this value and the detector area determines the noise equivalent power (*NEP*):

$$NEP = AH_n = \frac{V_n}{(\Delta f)^{1/2} R} \ (W/Hz^{1/2})$$ $\qquad\qquad$ **(6.4)**

The best photodetectors are characterized by the lowest values of *NEP*. However, the reverse parameter called a detectivity D is practically used::

$$D = \frac{1}{NEP} = \frac{R(\Delta f)^{1/2}}{V_n}$$ $\qquad\qquad$ **(6.5)**

It was shown that

$DA^{1/2} = \text{const}$ $\qquad\qquad$ **(6.6)**

The special detectivity relates to the frequency interval 1 Hz and the unit detector area 1 cm^2:

$$D^* = \frac{A^{1/2}}{NEP} = \frac{R(A\Delta f)}{V_n} \ (cm\ Hz^{1/2}W^{-1})$$ $\qquad\qquad$ **(6.7)**

The time constant τ is bound with the carriers lifetime in the semiconductor base of the photodetector and characterizes the reaction speed of the device. This parameter is defined by different recombination processes in the semiconductor.

The principal recombination processes in $Cd_xHg_{1-x}Te$-based photodetectors are Auger-recombination and the recombination on local centers.

10. PROPERTIES OF THE PHOTODETECTORS

HgCdTe material (both bulk and thin films) remains one of the most important compounds for producing novel IR-optoelectronic devices. Among them Auger-Suppressed HgCdTe Infrared Photodiodes which are basically p-i-n structures [37]. Their operation is not the same as for typical silicon *p-i-n*-photodiodes. First, due to limitations in HgCdTe technology [37], the middle 'i' region is not actually undoped, but low-doped. One refers to it as lightly doped '*v*' (*n*-type) or 'π' (*p*-type). This is because the controlled doping of HgCdTe has to be at a higher level than the background doping due to native defects and impurities. The minimum controllable doping that can be achieved is on the order to mid-10^{14} cm^{-3} [37]. Then, the intrinsic carrier concentration in HgCdTe is very high as the temperature is increased. It can reach values as high as 10^{15} cm^{-3}, or even 10^{16} cm^{-3}, *i.e.* values higher than the actual extrinsic doping level. Therefore, since Auger processes are directly proportional to the number of electrons and holes available, they result in large detector noise in HgCdTe IR devices operated at high temperatures where the absorber layer is intrinsic at equiblibrium. The $P+/\pi$ or $N+/v$ heterojunction is referred to as the exclusion junction [37]. The N^+/π or P^+/v heterojunction is referred to as the extraction junction. For example, we consider a $P^+/v/N^+$ structure operated at high temperature so that the '*v*' absorber region is intrinsic in thermal equilibrium (Fig. (**31**)) [37]. In this case, the intrinsic carrier concentration is in the mid-10^{15} cm^{-3} for T=280 K. When HgCdTe is highly *n*-type doped like in the N^+ region, the Fermi level lies in the conduction band. The energy required to generate *e-h* pairs by Auger generation is therefore increased to a value higher than the energy band gap, hence hole generation is decreased in the N^+ region. Now, when applying an increasing reverse bias, the P^+/v extraction junction extracts the minority holes from the '*v*' absorber region faster than they can be injected from the N^+ region. Therefore, the hole concentration drops significantly, below its thermal equilibrium value, in the absorber *v* region. The electron concentration is then also reduced below the equilibrium value in order to maintain charge neutrality. Analysis of current-voltage characteristics and calculation of net Auger recombination in the device showed that high sensitivity at high temperatures can potentially be achieved [37].

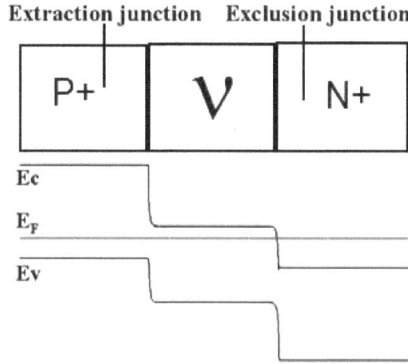

Figure 31: HgCdTe high operating temperature infrared photodiode structure and its corresponding calculated energy band profile at zero bias [37].

Now we briefly consider properties of various photo-devices based on $Cd_xHg_{1-x}Te$ proposed for different atmospheric windows.

Photoresistors. The main parameters are voltage sensitivity, spectral range of sensitivity, and detectivity. Expression (10.1) is valid at the following conditions: weak signal, equality of electrons and holes lifetimes, and low surface recombination for an open circuit consisting of consequently connected photoresistors, load resistor, and the bias voltage [38]:

$$R_v = \frac{\eta\lambda}{hc}\frac{V\tau}{lwtn(1+\omega^2\tau^2)^{1/2}} \tag{7.1}$$

where η is a quantum efficiency, V is an applied bias, τ, n are a lifetime and concentration of electrons, respectively, ω is a modulation frequency, l,w,t are length, width, and thickness of the photoresistor.

The detectivity under its by generation-recombination noises and considerably high applied voltage is as follows:

$$D^* = \frac{\eta\lambda}{2hc}\left[\left(\frac{n+p}{np}\right)\frac{\tau}{t}\right]^{1/2} \tag{7.2}$$

A maximum value of the detectivity is [38],

$$D^* = 0.45 \left[\frac{(n+p)\alpha\tau}{np} \right]^{1/2} \frac{\lambda}{hc} \qquad (7.3)$$

where α is an absorption coefficient, n, p are elctron and hole concentrations, respectively. For an intrinsic material the detectivity is determined as,

$$D^* = 0.64 \frac{\lambda}{hc} \left(\frac{\alpha_i \tau_i}{n_i} \right)^{1/2} \qquad (7.4)$$

In the most cases the detector as well as the radiation of the origin which is detected are in the atmosphere at the temperature about 300 K. Therefore, the detector registers not only the so-called useful radiation (*i.e.* the radiation of the origin), but also a background radiation. The backgroud has a radiation spectrum of the black body at ~300 K. Thus, the detector should function under conditions providing the noise level lower or equal to the level of noise caused by fluctuations of the number of photons incident on the working surface of the detector these conditions are called BLIP, Background Limited Infrared Photodetector). At these conditions the fluctuation of the incident photons number defines the minimum number of photons supplied by the object of detectivity. It allows separating the object signal from the photon noise of the background.

The background radiation can be negligible if the concentration of charge carriers generated by the radiation is about the same order of magnitude as their equilibrium concentration. The total concentration of the charge carriers is as follows:

$$n = n_o + \Delta n = n_o + \eta\Phi_B \frac{\tau}{t}, \qquad (7.5)$$

$$p = p_o + \Delta p = p_o + \eta\Phi_B \frac{\tau}{t}, \qquad (7.6)$$

where n_o and p_o are equilibrium concentrations of electrons and holes, η is a quantum efficiency (a number of electron-hole pairs generated by the absorption of one photon), Φ_B is a flow density of the background photons, τ is the carriers lifetime, and t is a thickness of the photoresistor.

Substituting (10.6) in (10.2) at $\Delta n = \Delta p \gg n_o$, p_o gives:

$$D^*_{BLIP} = \frac{\lambda}{2^{1/2} hc}\left(\frac{\eta}{\Phi_B}\right) \tag{7.7}$$

At significant surface recombination the equation (10.2) is to be modified: the carrier lifetime for a bulk active element is replaced with effective lifetime of charge carriers for thin samples ($t<<L$) [64]

$$\tau_{ef} = \left[\frac{1}{\tau} + \frac{s_1 + s_2}{t}\right]^{-1} \tag{7.8}$$

Here s_1 and s_2 are recombination rates on the illuminated and dark surfaces of the device. If $\tau >> t/(s_1 + s_2)$, we have according to (10.2):

$$D^* = \frac{\eta\lambda}{2hcn_i}\left(\frac{n+p}{s_1+s_2}\right)^{1/2} \tag{7.9}$$

The photodetectors based on solid solutions HgCdTe experience degradation of their working functions under storage at the room temperature. This degradation is caused by mercury evaporation which can even lead to change of the type of conductivity of the active element. ZnS protective coatings are appeared as one of the most effective protection means for HgCdTe-based photodetectors.

Photodiodes. The photovoltaic effect was first observed on bulk HgCdTe samples at the beginning of the era of the narrow gap solid solutions [1]. The structures are diffusion junctions and implanted active elements.

If there is a potential barrier and an internal electric field in a semiconductor and this field separates the electron-hole pairs generated by the photons, the photovoltaic effect appears. It means that the photovoltage (10.10) is observed at the electrodes of the device under radition:

$$V_{ph} = I_{ph} R \tag{7.10}$$

Here $I_{ph} = \eta q N_\lambda$ is a short-circuit current of the photodiode, R is a differential resistance of the photodiode. Thus, the voltage detectivity of the device is

$$R_V = \frac{\lambda q}{hc} \eta R \qquad (7.11)$$

The detectivity of the photodiode is written as:

$$D^*_\lambda = \frac{\lambda}{hc} \left(\frac{\eta}{2\Phi_B} \right)^{1/2} \qquad (7.12)$$

The main parameter of the photodiodes is a high performance. The restriction of the device performance is caused by the time of the carriers diffusion from the generation point to the space charge region, and by the time constant bound with the capacity C of *p-n*-(or hetero)junction ($\tau_{RC} = RC$).

The time of carriers diffusion at the distance *t* from the generation point to the junction at *t<<L*, where *L* is a diffusion distance, is equal to $\tau_D = t/2D$, *D* is the diffusion coefficient of minority charge carriers. The considerably high perfomance is typical for *p-n*-junctions built in the structures with graded element composition [23, 27-29, 57-64].

ACKNOWLEDGEMENTS

Declared none.

CONFLICT OF INTEREST

The author(s) confirm that this chapter content has no conflict of interest.

REFERENCES

[1]   Lawson, W. D.; Nielsen, S.; Putley E. H.; Joung A. S. Preparation and properties of HgTe and mixed crystals of HgTe-CdTe. *J. Phys. Chem. Sol.*, **1959**, *9*, 325-329.
[2]   Schmit, J. L.; Stelzer, E.L. Temperature and alloy compositional dependences of the energy gap of $Cd_xHg_{1-x}Te$. *J. Appl. Phys.*, **1969**, *40(12)*, 4865-4869.
[3]   Wiley, J. D.; Dexter, R. N. Helicons and nonresonant cyclotron absorption in semiconductors. *Phys. Rev.*, **1969**, *181*, 1181-1190.
[4]   Scott M. W. Electron mobility in $Hg_{1-x}Cd_xTe$. *J. Appl. Phys.*, **1972**, *43*, 1055-1062. Long, D.; Schmit, J. L. *Semiconductors and Semimetals*, v. 5; Academic Press: New York, **1970**.
[5]   Cruceanu, E.; Niculescu, D.; Nistor, N.; Aldea, A.; Investigation of pseudobinary system ZnTe-HgTe. *Rev. Roum. Phys.*, **1964**, *9(5)*, 499-506.

[6]　Shneider, A.D.; Tsiutsiura, D. I.; Makarenko, V.V.; Grigorovich, G. M. Some electrical and photoelectric properties of the system HgTe-ZnTe. *Ukr. Fiz. Zh.*, **1965**, 10(8), 915-917.

[7]　Sher, A., Chen, A.-B.; Spicer W. E.; Shih. Effects influencing the structural integrity of semiconductors and their alloys, *J. Vac. Aci. Technol. A*, **1985**, 3(1), 105-111.

[8]　Toulouse, B. ; Granger, R. ; Rolland, S.; Triboulet R. Band-gap in $Hg_{1-x}Zn_xTe$ solid solutions. *J. Physique*, **1987**, *48*, 247-252.

[9]　Shneider, A. D.; Tsiutsiura, D. I. Electrophysical properties of $Zn_xHg_{1-x}Te$ alloys. *Phys. Stat. Sol. (a)*, **1971**, *5*, K39-K41.

[10]　Sydorchuk, P. G. Epitaxial crystallization of $Zn_xCd_yHg_{1-x-y}Te$ from melts. *Inorganic Mater. (Russia)*, **1979**, *15(9)*, 1538-1542.

[11]　Sydorchuk, P. In: *Prace ITME*, Warszawa, **2000**; pp. 198-202.

[12]　Cebulski, J.; Kakol, T.; Polit, J.; Sheregii, E. M.; Sydorchuk P. G. Peculiarities of the Magnetotransport in Quaternary $Zn_xCd_yHg_{1-x-y}Te$. In: Proceedings of NGS10IPAP Conf. Series, pp. 76-79.

[13]　Sydorchuk, P.; Khlyap, G.; Andrukhiv, A. Growth and some properties of heterostructures based on new narrow-gap semiconductor ZnCdHgTe. *Cryst. Res. Technol.*, **2001**, *36(4-5)*, 361-369.

[14]　Kriessl, O. PhD Thesis. Efficient numerical simulation of liquid phase epitaxy. University of Augsburg, 2005.

[15]　Khlyap G.; Sydorchuk, P. Growth and electrical properties of new semiconductor compound ZnCdHgTe. *Cryst. Res. Technol.*, **2001**, *36(8-10)*, 1027-1034.

[16]　Khlyap, G.; Sydorchuk, P.; Polit, J.; Oszwaldowsky, M. Investigation of ZnCdHgTe thin films and ZnCdHgTe-based heterostructures, *J. Cryst. Growth*, **2005**, *275*, e927-e930.

[17]　Shinohara, K.; Ueda, R.; Ohtsuki, O; Ueda, Y. The crystal growth of homogeneous $Hg_{1-x}Cd_x$ Te by the Bridgman method. *Japan. J. Appl. Phys.*, **1972**, *11*, 273-274.

[18]　Dziuba E.Z., Preparation of $Cd_x Hg_{1-x}$ Te crystals by the vertical-zone melting method. *J. Electrochem. Soc.* **1969**, *104*. 104-106.

[19]　Ueda, R.; Ohtsuki, O.; Shinohara, K.; Ueda, Y. Crystal growth of $Hg_{1-x}Cd_xTe$ using Te as a solvent. *J. Cryst. Growth*, **1972**, *13/14*, 668-671.

[20]　Gritsenko, Yu.I. In: *Izvestiya AN USSR (Russia), Series Physics*, **1957**, *21(1)*, 154-157.

[21]　Parker S. G., Kraus H., Method of producing homogeneous ingots of mercury cadmium telluride. US3468363. **1969**.

[22]　Johnson R. E. Method for preparing single crystal pseudobinary alloys. US3622399. **1971**.

[23]　Cohen-Solal, G. PhD Thesis. University of Paris, **1967**. *See also*: Sydorchuk, P.G. PhD Thesis.Investigation of crystallization conditions influence on physical properties of crystals and epitaxial structures of mercury-cadmium tellurides. Physical-technical institute named after A.F.Ioffe, St-Petersbourg, 1978.

[24]　Petty, M. C.; Juhasz, C.; Optical absorption in $Zn_xHg_{1-x}Te$ thin films. *J. Phys. D: Appl. Phys.*, **1976**, *9*, 2305-2316.

[25]　Kaiser, D. L.; Becla, P. *In: Proceedings of Materials Research Society (USA)*, **1987**, *90*, 397-404.

[26]　Nelson, H. Epitaxial growth from the liquid state and its application to the fabrication of tunnel and laser diodes. *RCA Rev.*, **1963**, *24(4)*. 603.

[27]　Konnikov, S.G.; Ogorodnikov, V. K.; Sydorchuk P.G. CdTe-$Cd_xHg_{1-x}Te$ heterostructures. *Phys. Stat. Sol.(a)*, **1975**, *27(1)*, 43-48.

[28] Maciolek R. B.; Speerschneider C. J. Growth of mercury cadmium telluride by liquid phase epitaxy. US3902924. **1975**.

[29] Ivanov-Omskii V.I.; Ogorodnikov V.K.; Sydorchuk P.G. In: *Growth and doping of semiconductor crystals and films*, Nauka Publishers: Novosibirsk, **1977**, Part *1*, pp.141-144.

[30] Wang, C. C.; Shin, S. H.; Chu, M.; Lanir, M.; Vanderwyck, A. H. B. Liquid phase growth of HgCdTe epitaxial layers. *J. Electrochem. Soc.*, **1980**, *127(1)*, 175-179.

[31] Lanir, M.,; Wang, C. C.; Vanderwyck, A. H. B., Backside-illuminated HgCdTe/CdTe photodiodes. *Appl. Phys. Lett.*, **1979**, *34*, p. 50-52.

[32] Sher, A.; Eger, D.; Zemel, A. Mercury zinc telluride, a new narrow-gap semiconductor, *Appl. Phys. Lett.* **1985**, 46(1), 59-61.

[33] Faurie, J. P. Developments and trends in MBE of II-VI Hg-based compounds. *J. Cryst. Growth*, **1987**, *81*, 483-486.

[34] Berding, M.A.; Krishnamurthi, S.; Sher, A.; Chen A.-B. Electronic and transport properties of HgCdTe and HgZnTe. *J.Vac.Sci.Technol.A5*, **1987**, *No.5*, p.3014-3018.

[35] Faurie, J. P.; Boukerche, M.; Reno, J.; Hsu, C. Molecular beam epitaxy of alloys and superlattices involving mercury. *J.Vac.Sci.Technol. A3*, **1985,** p. 55-59

[36] Emelie, P.-Y. PhD Thesis. HgCdTe Auger-Suppressed Infrared Detectors Under Non-Equilibrium Operation. University of Michigan, **2009**.

[37] Sivananthan, S.; Chu, X.; Boukerche, M.; Faurie J. P. Growth of Hg1-xZnxTe by molecular beam epitaxy on a GaAs(100) substrate. *Appl. Phys. Lett.*, **1985**, *47*, 1291-1293.

[38] Schofield, K. H. Method of laser ablation of semiconductor structures. US5932485. **1999**.

[39] Oszwaldowski, M.; Berus, T.; Rzeszutek, J.; Sydorchuk, P. *et al.* Pulsed laser deposition of II-VI semiconductor thin films and their layered structures. *J.Alloys&Compounds*, **2004**, *371*, 164-167.

[40] Qingxue, W.; Jianrong, Y.; Yanfeng, W.; Weizheng, F.; Li, H. In: Proceedings SPIE, SPIE: Bellingham (WA), **2005**; Vol. *5640.*, pp. 629-636.

[41] Zandi, B.; Vasileska, D.; Wijewarnasuriya, P. Modeling Mercury Cadmium Telluride (HgCdTe) Photodiodes. *Army Research Laboratory, Report ARL-TR-5033*, **2009**, pp.1-10.

[42] Dvoretsky, S. A.; Ikusov, D. G.; Kvon, Z. D.; Mikhailov, N. N.; Remesnik, V. G.; Smirnov, R. N.; Sidorov, Yu. G.; Shvets, V. A. HgCdTe quantum wells grown by molecular beam epitaxy. *Semiconductor Physics, Quantum Electronics & Optoelectronics,* **2007**. *10(4)*, 47-53.

[43] Kinch, M. HgCdTe: Recent Trends in the Ultimate IR Semiconductor. *Journal of Electronic Materials,* **2010**, *39(7)*, 1043-1052.

[44] Benson, D.J.; Dinan, J. H.; Martinka, M.; Almeida, L.A.; Boyd, P.R.; Andrew, Jr.; Stoltz J.; Kaleczyc, A. W. Fullerene addition in photoresist *via* incorporation in the developer. US20040185387. **2004**.

[45] Suzuki, K.; Hirano, R.; Kurita, H. Semiconductor substrate for epitaxial for epitaxial growth and manufacturing method thereof. US20090269271. **2009**.

[46] Nikolic, R.J.; Conway, A.M.; Nelson, A.J.; Payne, S. A. Energy resolution in semiconductor gamma radiation detectors using heterojunctions and methods of use and preparation thereof. US20090294680. **2009**.

[47] Khlyap, G.; Sydorchuk, P. In: *Proceedings of Int. Conf. On Advanced Semicond. Dev. and Modeling ASDAM'02*, **2002**, 219-223.

[48]   Khlyap, G.; Sydorchuk, P. In: *Proceedings of Int. Conf. On Advanced Semicond. Dev. and Modeling ASDAM'02*, **2002**, 99-103.

[49]   Bonch-Bruevich, V.L.; Zvyagin, I.P.; Keiper, R.; Mironov, A.G.; Enderlein, R.; Esser, B. *Electronic Theory of Disordered Semiconductors*. Nauka Publishers: Moscow, **1981**.

[50]   Krowne, C.M. Semiconductor heterostructure nonlinear Poisson equation. *J. Appl. Phys.*, 1989, *65(4)*, 1602-1620.

[51]   Khlyap, H.M.; Sydorchuk, P.G.; Andrukhiv, A.M. Recent patents on semiconductor thin film technology. *Recent Patents on Materials Science*, **2011**, *4*, 50-55.

[52]   Raikh, M.E.; Ruzin, I.M.; Shklovskii, B.I.; Effect of localized states in the barrier on the fluctuation tunneling current through the contact metal-semiconductor. *Semiconductors (Russia)*, **1988**, *22*, 1979-85.

[53]   Grekhov, I.V.; Shulekin, A.F.; Veksler, M.I.; Mechanism of steady-state reverse current-voltage characteristics of the MDS-structures with charge transfer. *Semiconductors (Russia)*, **1995**; *29, 229-34*.

[54]   Bychkovskii, D.N.; Konstantinov, O.V. Effect of the built-in charge in an isotype heterojunction on capacitance-voltage characteristics of the barrier structure. *Semiconductors (Russia)*, **1992**; *26*, 921-6.

[55]   Kal'fa, A.A.; Chikin, V.V. Current-voltage characteristics of the Schottky diodes with lightly doped semiconductor layer in the space-charge region. *Semiconductors (Russia)*, **1992**; *26*, 1024-7.

[56]   Belyaev, A.P. About current-voltage characteristic of the heterostructure based on disordered gallium arsenide. *Semiconductors (Russia)*, **1995**; *29*, 70-77.

[57]   Piotrowski, J.; Rogalski, A. Polprzewodnikowe detektory podczerwieni. WNT: Warszawa, **1985**.

[58]   Ivanov-Omskii V.I.; Kolomiets, B.T.; Ogorodnikov, V.K.; Sydorchuk, P.G. USSR Patent Application № 481087, 1975 г.

[59]   Ovsyuk, V.N.; Sidorov Yu.G.; Vasil'ev V.V., Shashkin V.V. Matrix photodetectors 128x128 based on HgCdTe layers and multilayered heterostructures with GaAs/AlGaAs quantum wells. *Applied Physics (Russia)*, **2000**, **No.5**, 70-79.

[60]   Rogalski, A. New trends in development of electromagnetic radiation detectors, *Postepy Fizyki*, **2000**, *51(2)*, 57-68.

[61]   Tsiutsiura, D.I.; Sydorchuk, P.G.; Grigorovich, G.M. Preparation and some properties of *p*-*n*-junctions based on $Zn_xHg_{1-x}Te$. *Semiconductors (Russia)*, **1970**, *4*, 1567-1569.

[62]   Lloyd, J. M. Thermal imaging systems. Plenum Press: New York-London, **1975**

[63]   Kinch, M. A.; Borrello, S. R. 0.1 eV HgCdTe photodetectors. *Infrared Phys.*, **1975**, *15*, 111-124.

[64]   Dmitriev, E.P. Working-out of elemental base of videoinformation phodetectors. *Electronics: Science, Technology, Bysiness (Russia)*, **2005**, *No.2*, 74-79.

Send Orders of Reprints at reprints@benthamscience.net

CHAPTER 2

High-Effective Solar Cells, Thermodynamics and Physics of the Universe

Viktor I. Laptev [1,*] **and Halyna Khlyap** [2]

[1]*Russian New University, Radiostr.22, RU-105005 Moscow, Russian Federation and* [2]*Distelstr.11, D-67657 Kaiserslautern, Germany*

Abstract: Conventional silicon-based solar cells production accounts principal characteristics of the semiconductor compounds presenting the main building part of the device. The first part of this e-lecture reports unique results on fabrication and quite uncommon performance of Si-based solar cells with novel contacts chemically deposited on traditionally prepared semiconductor part of the solar cell. How does the high-effective solar radiation conversion occur and what thermodynamics aspects are to be involved in theoretical explanation and further practical applications? The next part of the e-lecture proposes answers. The thermodynamics is an effective tool used in the physics of the Universe. Photonic condensate and the relic radiation are not only exotic objects for theoretical astrophysics but also present possible power tools for achieving high-effective energy conversion. This question is considered in further parts of the e-lecture.

Keywords: Conventional silicon solar cell, metallic nanoclusters, thermodynamics, solar energy conversion, antenna states, relic radiation, photonic condensate, solar radiation, a green plant.

1. INTRODUCTION

Silicon (bulk and thin films, micro- and nanocrystalline Si) and other nanostructural semiconductors are materials of choice for producing high-effective solar cells. In this part we consider main fundamentals of solar cell design and some aspects of technology.

Fig. (**1**) plots a generalized current-voltage characteristic of conventional solar cells [1].

The main parameters of the solar cell as a device are following: I_{sc} and V_{oc} are short circuit current and open circuit voltage, respectively, I_m and V_m are the

---

*Address correspondence to Viktor I. Laptev: Russian New University, Radiostr.22, RU-105005 Moscow, Russian Federation; Tel: +49 7031 608113; E-mails: viktor.laptev@yahoo.com

current and voltage at optimal operation point, and P_m is the maximum achievable power output [1]. The performance of the solar cell is defined by the fill factor FF (the ratio of peak output ($V_m I_m$) to the variable ($V_{oc} I_{sc}$)) and efficiency η which is the ratio of the photovoltaically generated electric output of the cell to the luminous power falling on it [1]:

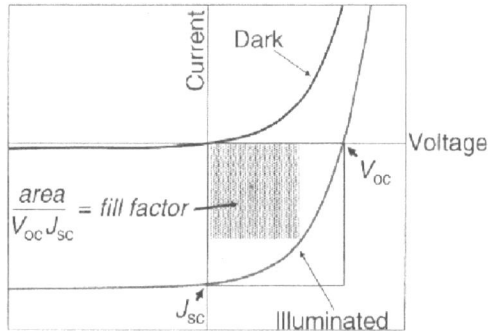

Figure 1: Current-voltage characteristic of an ideal solar cell [1].

$$FF = V_m I_m / V_{oc} I_{sc} \tag{1.1}$$

$$\eta = V_m I_m / P_{light} = FF\ V_{oc} I_{sc} / P_{light} \tag{1.2}$$

As we can see, the principal task of the solar cell technology is increasing the efficiency of the device. Solar cells based on semiconductor and metallic nanostructures successfully fabricated and show good perspectives. Authors [2] have proposed a nanostructured photovoltaic device based on the ZnO nanostructures/poly(3-hexylthiophene) (P3HT):TiO$_2$ nanorod hybrid by solution processes at low temperature. An array of ZnO nanorods with a larger size of ~50 nm in diameter and ~180 nm in length are grown to provide direct pathways for efficient charge collection. TiO$_2$ nanorods with a size of ~5 nm in diameter and ~20–30 nm in length are incorporated into polymers to facilitate charge separation and transport by providing an increased interfacial area and a more effective transport pathway. The device performance with the inclusion of TiO$_2$ nanorods exhibits a seven times increase in the short circuit current with respect to that without TiO$_2$ nanorods. The device performance can be further enhanced after completely removing the residual surfactant on the TiO$_2$ nanorods using the ligand exchange method, giving a short circuit current density of 2.67 mA cm^{-2}

and a power conversion efficiency of 0.59% under Air Mass 1.5 (100 mW cm^{-2}) illumination [2].

A novel Dye-Sensitized Solar Cell (DSSC) scheme for better solar conversion efficiency is proposed in [3]. The distinctive characteristic of this novel scheme is that the conventional thin film electrode is replaced by a 3D nanostructured indium tin oxide (ITO) electrode, which was fabricated using RF magnetron sputtering with an anodic aluminum oxide (AAO) template. The template was prepared by immersing the barrier-layer side of an AAO film into a 30wt% phosphoric acid solution to produce a contrasting surface. RF magnetron sputtering was then used to deposit a 3D nanostructured ITO thin film on the template. The crystallinity and conductivity of the 3D ITO films were further enhanced by annealing. Titanium dioxide nanoparticles were electrophoretically deposited on the 3D ITO film after which the proposed DSSC was formed by filling vacant spaces in the 3D nanostructured ITO electrode with dye. The measured solar conversion efficiency of the device was 0.125%. It presents a 5-fold improvement over that of conventional spin-coated TiO2 film electrode DSSCs [3].

Authors [4] report the dark and illuminated I-V curves and spectral response characteristics of six-stack InAs/InGaAs quantum dots-in-a-well (DWELL) solar cells. The short circuit current density, open circuit voltage, and external quantum efficiency of these cells under air mass 1.5G at 100 mW/cm^2 illumination are presented and compared with a GaAs control cell. The InAs DWELL cells show higher short circuit density and better efficiency compared to the control cells, confirmed by spectral response measurements. It is found that the trend in the quantum dot (QD) cell ideality factor with voltage is opposite that of the control cell making the quantum dot devices attractive for high concentration. By comparing the dark current density for the QD cell and GaAs control cell, we have conservatively estimated the concentration level at which the QD solar cells would surpass the open circuit voltage of conventional GaAs devices [4].

Effects of very high frequency- plasma enhanced chemical vapor deposition (VHFPECVD) using diluted ultrapure silane at higher dilution ratio (R>30) on microstructures and optical characteristics of hydrogenated nanocrystalline silicon (nc-Si:H) film were studied in [5]. Nanocrystalline silicon films were prepared by magnetron sputtering at RF power ranging from 50 to 300 W. It was found that the transition from amorphous phase to nanocrystalline phase occurred between 100 W and 150 W. The nucleation mechanism toward nc-Si:H near the transition point of amorphous phase was discussed based on transmission electron microscopy with atomic scale. Further, it is suggested from UV-visible spectroscopy that nc-Si:H films with the best optical properties would be obtained near the transition point from the amorphous phase to the crystalline phase [5].

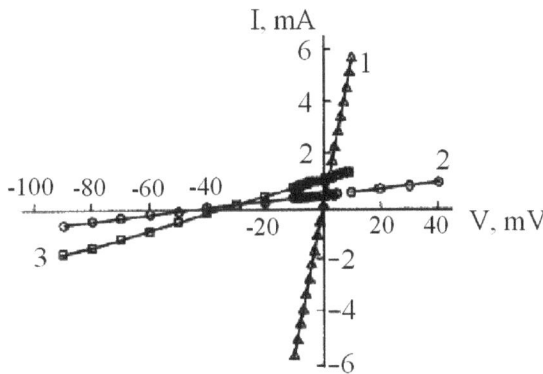

Figure 2: Electrical properties of: (1) a silver contact strip, (2) a contact strip with Cu clusters in silver pores, and (3) a strip with a Cu-layer on the surface and Cu clusters in Ag pores [6].

Metallic nanoclusters as a principal part of the solar cell also submit into increase of their efficiency [6]. Electrical properties of copper clusters in porous silver contacts of silicon solar cells have been investigated and showed the dark current unknown in metals [6]. The copper clusters have been introduced into the pores of the silver strips by chemical deposition. The resistivity of the silver contact strips was measured at room temperature with Keithley 236 source-measure unit. The current-voltage characteristics of the contacts are shown in Fig. (**2**) [6].

The contact strips without copper clusters demonstrated metallic conductance. However, the lines for the contacts with copper clusters do not cross the origin of the current-voltage coordinates for both forward and reverse currents. A current through a metal in absence of an external electric field is observed (Fig. (**3**)). The experiment had been performed during 10 min. The dark current in the Ag/Cu-contact on the illuminated side of a silicon solar cell is up to 5% of the rated value. The efficiency of solar cells after the copper deposition had not changed or increased up to 1 %. [6].

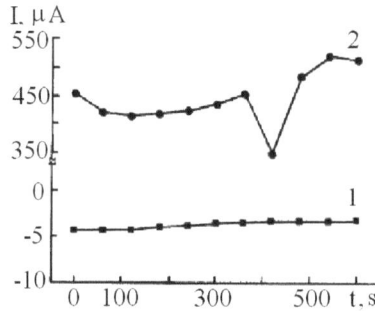

Figure 3: Time dependence of the (1) dark and (2) light currents at zero applied bias in contact strips with Cu clusters in silver pores [6].

Authors [7] have demonstrated that Ag nanocrystals made by chemical synthesis have been used in solar cell applications as a part of light trapping. The shape, crystal structure, defects and composition of these nanocrystals have been studied in detail. Samples with different ratios of silver solution (AgNO$_3$) and reductant (NaBH$_4$) were made, and a difference in nanocrystal size was observed. High resolution transmission electron microscopy (HRTEM) and diffraction patterns showed that the samples contained mostly Ag nanocrystals, and some of them contained Ag$_2$O nanocrystals as well. Some nanocrystals contained large defects, mostly twinning, which induced facets on the nanocrystal surface [7].

Carriers transport in solar cells based on various semiconductors is of special interest. As reported in [8], three-dimensional solid state solar cells, or 3D-cells, are based on nanostructured n-type TiO$_2$ electrodes as known from dye-sensitized solar cells, and completed with a p-type solid state semiconductor absorber CuInS$_2$ which is absorber and hole conductor in one. To study the electron transport in this specific type of nanostructured solar cells, they are subjected to intensity modulated photocurrent spectroscopy. The photocurrent response of the 3D-cell turns out to be four orders of magnitude faster than the response of nanostructured dye-sensitized solar cells. No voltage dependency is found for the time constants, so we conclude that there is no macroscopic electric field present in the 3D-cell and that diffusion is the major driving force for carrier transport in the 3D-cell. The fact that an increase in illumination intensity results in a decrease in time constant, supports this idea because it shows, that charging of the nanostructured TiO_2 film due to trap filling enhances the diffusion. The estimated diffusion coefficient of electrons in the nanostructured TiO_2 is in the order of 10^{-8} m^2/s for high illumination intensities [8].

2. THERMODYNAMICS OF SOLAR ENERGY CONVERSION [6]

Any solar cell can be considered as a heat engine. The process of *light→electricity* conversion is described by the thermodynamic laws. This chapter illustrates how the thermodynamics is applied for the subject.

It is known [6] that the efficiency of the photon re-emission for a black body at the temperature T_A and the radiation with temperature T_S is,

$$\eta_U = 1 - (T_A/T_S)^4 \tag{2.1}$$

The work between these limit temperatures produced as the Carnot cycle runs is performed with efficiency:

$$\eta_C = 1 - T_A/T_S \tag{2.2}$$

The work of the Carnot cycle running in the temperature range from the $T_{\text{black body}}$ to the temperature T_0 of heat receiver is performed with an efficiency:

$$\eta_0 = 1 - T_0/T_A \tag{2.3}$$

The values $\eta_C\eta_U$ and $\eta_0\eta_C\eta_U$ correspond to the product of the reversible cycle efficiencies η_0, η_C and the efficiency η_U of irreversible photon re-emission without work production. One should note that the efficiency of solar energy conversion in the reversible cycle and reversible photon re-emission is [6]:

$$\eta_L = 1 - (T_A/T_S)^4 - 4T_0 [1 - (T_A/T_S)^3]/3T_S \tag{2.4}$$

The following method for calculating relative contributions of reversible re-emission is proposed: the temperature dependence of η_L from equation (2.4) is shown by line "LB" in Fig. (**1**). We also take into account that this line is (by definition) a graphic illustration of the reversible processes only. Then the η-coordinates of the points on the lines representing the combinations of efficiencies of the reversible and irreversible processes (for example, $\eta_C \eta_U$, $\eta_0 \eta_U$, $\eta_0 \eta_C \eta_U$ or lines "KB", "CEB" and "CDB" in Fig. (**1**)) are proportional to their shares.

For example, let the point a in Fig. (**1**) denote the conversion of solar energy along the line "CE" with efficiency of 30%. Let us draw an isotherm crossing the point a; the intersections of this line with the line "LB" and the y-axis give us the values $\eta_L = 0.93$ and $T_A = 430$ K. Then, according to the lever principle, the point a corresponds to $(\eta_L - 30)100/\eta_L = 68\%$ of irreversible processes (photon re-emission) and 32% of reversible ones (Carnot cycle).

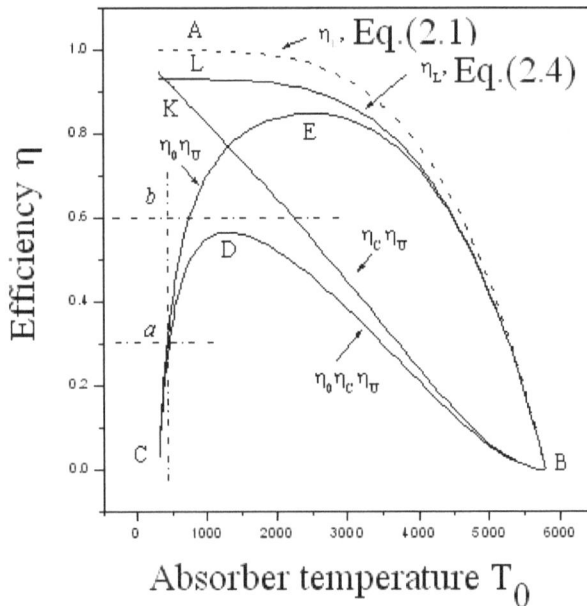

Figure 4: Comparison of thermodynamic efficiency limitations for the solar energy conversion with and without entropy production [6].

The zone theory proposes mechanisms of converting solar energy into work with efficiency of 43% and 60% [6]. Let us denote the last value by the point b on the isotherm in Fig. (**1**). According to the lever principle, point b corresponds to reversible and irreversible photon re-emission processes without work production

(η=50% for both processes). It is shown in Fig. (**1**) that the threshold of 30% efficiency early discovered by W. Shockley and H. Queisser can be overcome only if photons are re-emitted reversibly without work production together with the conversion of solar heat into work.

There are no theoretical restrictions for achieving efficiency of η_L equal to 93% at the room temperatures (line "LB" in Fig. (**1**)). It can be reached if the problem of separating the photon re-emission and the work production processes is solved and the way of transformation of irreversible photon re-emission into reversible one is to be found.

3. SOLAR RADIATION → USEFUL WORK CONVERSION [6]

The contemporary thermodynamic descriptions of *solar heat → work* conversion are proposed as follows: i) one should draw up a balance sheet of energy and entropy flows [6]; ii) we have to use the method of cycles in order to give a visual description of the solution of balance equations [6]. The paths along which the energy exchange between radiation and matter is accompanied by the performance of a maximum work are to be found. The work reaches its maximum when a quasi-static process takes place in a device. However, substance and radiation are not in equilibrium in solar cells and quasi-static conversion of the solar energy is not achieved.

For this reason, the optimal configurations for various irreversible thermodynamic engines are determined using, for example, the method of endoreversible thermodynamics of solar energy conversion [6]. Endoreversible engines are irreversible engines where all irreversibilities are restricted to the coupling of the engine to the external world [6]. It is assumed that the internal reversible part of endoreversible engine is a Carnot cycle.

It is proposed to look for a continuous series of equilibrium states outside the irreversible or endoreversible engine, to isolate these states into separate processes, and to use them to obtain a higher efficiency for general non-quasi-static solar energy conversion. We can consider those equilibrium processes as a base of the "exoreversible" additional device for the irreversible or endoreversible engine.

In this lecture we show how to combine principles of thermodynamics and quantum theory for the cyclic processes.

3.1. Theoretical Model

We use the well-known model of a solar energy converter shown in Fig. (**1**) [6].

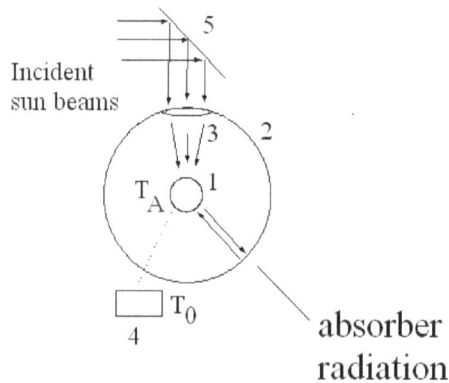

Figure 5: Model for solar energy conversion: 1 – black body, 2 – spherical cavity, 3 – lens, 4 – heat receiver, 5 – movable mirror [6].

The heat radiation absorber (1) is a black body under temperature T_A which is placed in the center of the spherical cavity (2) with mirror walls and a lens (3) making it possible to use optical methods in order to bring about a maximum concentration of radiation at the black surface. The heat receiver (4) under temperature T_0 (less than T_A) is in contact with the black body. The mirror (5) regulates the penetration of solar radiation into cavity (2). If the mirror is positioned as shown in Fig. (**1**), the cavity contains radiation with temperatures T_A and T_S. If the mirror blocks the penetration of the solar light, the cavity contains only radiation from the black body 1. Other types of radiation in the cavity are not considered here. In this model, radiant energy is converted at limit temperatures T_0=300 K and T_S=5800 K. The black body has a temperature T_A=320 K.

3.2. Energy Exchange between Radiation and Matter

The solar radiation gathered in the cavity 2 (Fig. (**1**)) with a volume V has energy $U_S=\sigma V(T_S)^4$ and entropy $S_s=4\sigma V(T_S)^3/3$, where σ is the Stefan-Boltzmann constant. The black body absorbs radiation and irradiates with energy $U_A=\sigma V T_A^4$

into the cavity 2. If T_A=320 K, these energies are in a ratio of $U_S/U_A \approx 10^6$, while $S_S/S_A \approx 6 \times 10^3$. As the volumes of radiation are equal, the amount of evolved heat ΔQ is proportional to the difference $T_A^4 - T_S^4$ and is equal to the area under the isochore "st" on the entropy diagram drawn on the plane formed by the temperature (T) and entropy (S) axes (Fig. (**2**)). The ratio of heat ΔQ to the solar energy U_S entering the cavity is,

$$\eta_U = (U_S - U_A)/U_S = 1 - (T_A/T_S)^4 \tag{3.1}$$

In our model, $\eta_U \rightarrow 1$, for $(T_A/T_S)^4 \sim 10^{-5}$.

3.2.1. Work Produced by Radiation and Matter Under Carnot Cycles

The absorbed radiant heat is converted into work by Carnot cycles involving matter or radiation. One such cycle is the rectangle "abcd" in Fig. (**2**). The work is performed during this cycle between the limit temperatures T_0=300 K and T_A=320 K with efficiency:

$$\eta_0 = 1 - T_0/T_A = 0.0625 \tag{3.2}$$

Radiation performs work under Carnot cycle with efficiency larger than η_0. We show the radiation absorption as an entropy diagram (Fig. (**2**)) demonstrating an isothermal transfer of radiation from the volume V_2 of the cavity (state "s") to the volume V_1 of the black body (state "p"). One can even reduce the radiation to state "p*" (Fig. (**3**)). Radiation reaches heat equilibrium with the black body (state "e") from these points either through the adiabatic process "p*e" or through the isochoric process "pe".

Let us represent the emission of radiation as its transfer from the volume of the black body (state "e") to the volume V_2 of the cavity along the isotherm T_A (state "t") (Figs. (**6**) and (**7**)). As the radiation fills the cavity, it performs a work equal to the difference between the evolved and absorbed heat. The radiation performs a considerable work if it reaches state "t*" in Fig. (**3**). Our calculations show that at T_A=320 K the work is performed along the path "sp*et*" with efficiency

$$\eta_C = 1 - T_A/T_S = 0.945 \tag{3.3}$$

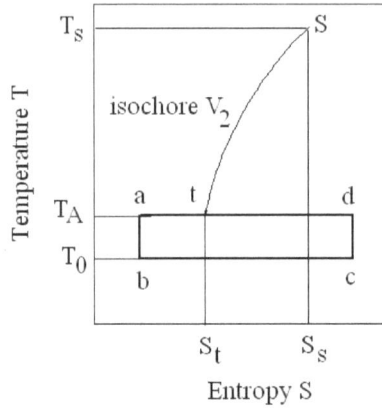

Figure 6: Entropy diagram showing isochoric cooling of radiation (line "st") in the cavity 2. The amount of evolved radiant heat is proportional to the area "sts_ts_s". The amount of heat converted into work is proportional to the area "abcd". The work is performed by matter in a thermal engine under Carnot cycle [6].

It is important to note that under returning radiation to its initial state "s" along the adiabatic line "t*s", it constitutes a Carnot cycle with the same efficiency η_C.

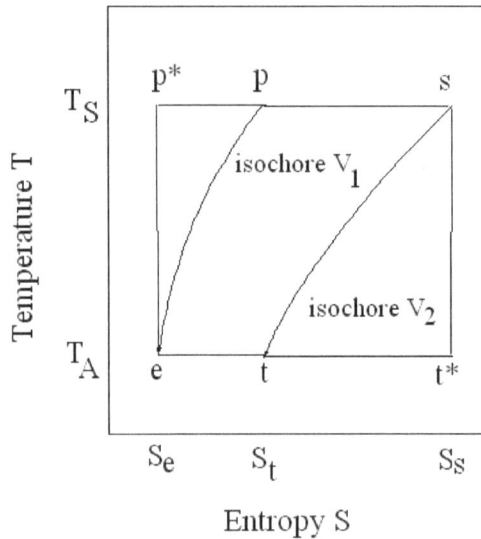

Figure 7: Entropy diagram showing some thermodynamic cycles for conversion of solar heat into work in the cavity 2 (Fig. (**1**)) with participation of the black body. Isotherms represent the absorption and emission of radiant energy. Lines "pe" and "p*e" correspond to cooling of radiation in the black body. Line "st" indicates the temperature and entropy of radiation in the cavity 2 [6].

3.2.2. Limits of the Conversion of Solar Heat into Work by Reversible Processes Other Than the Carnot Cycle [6]

Solar energy conversion results in combining different processes. Their mechanisms are mostly unknown. For this reason, one should attempt to find temperature dependence of the limiting efficiency for such a process (we consider the reversible one) (Fig. (**8**)) by balance equations describing energy and entropy flows. For solar radiation, it takes the form:

$$\eta_{AS} = 1 - 4T_A/3T_S + T_A^4/3T_S^4 \qquad (3.4)$$

For example, $\eta_{AS} = 0.926$ when $T_A=320$ K. The measure of magnitude of η_L is η_C, the efficiency of a Carnot cycle with the same temperature interval. The maximum difference $\eta_C-\eta_{AS}$ is about 18% when $T_A=3500$ K.

The efficiency of a reversible process under which radiation and matter perform work is equal to:

$$\eta_L = 1-(T_A/T_S)^4 - 4T_0 [1 - (T_A/T_S)^3]/3T_S \qquad (3.5)$$

For example, $\eta_L = 0.931$ if $T_A=320$ K and $T_0=300$ K.

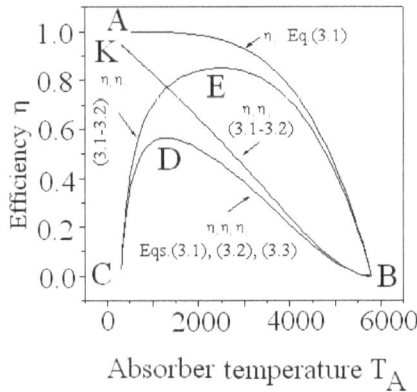

Figure 8: Comparison between efficiency of solar energy re-emission and efficiencies of conversion of solar heat into work. Line "AB" represents efficiency η_U of the reemissions, according to the Equation (3.1). Lines "KB", "CEB" and "CDB" show the limiting efficiencies of work performed in parallel with the solar energy reemission. Line "KB" includes the efficiency of Carnot cycle only for the radiation (Equation (3.3)), line "CEB" – only for matter (Equation (3.2)), line "CDB" – for matter and radiation in parallel [6].

3.2.3. Solar Energy Conversion as a Combination of Reversible and Irreversible Processes [6]

The absorption of radiation precedes the conversion of solar heat into work. In our model, the black body absorbs solar radiation and generates another radiation with a lower temperature. Heat is evolved in the process; it is either converted into work or irrevocably lost. For the sake of simplicity, let us assume that heat is lost with an efficiency of η_U from Equation (3.1). In Fig. (**3**) we show efficiencies for combinations of different Carnot cycles in the presence of irreversible losses of radiant energy. In these cases, the conversion efficiencies become smaller: $\eta_C\eta_U$ and $\eta_0\eta_U$. When energy is converted by two simultaneously running cycles, the efficiency becomes evenlower: $\eta_0\eta_C\eta_U$. In Fig. (**3**) the temperature dependence of these values is represented by lines "KB", "CE" and "CDB", respectively. In the region below each line, the work obtained is reconverted into heat by irreversible processes (Fig. (**8**)).

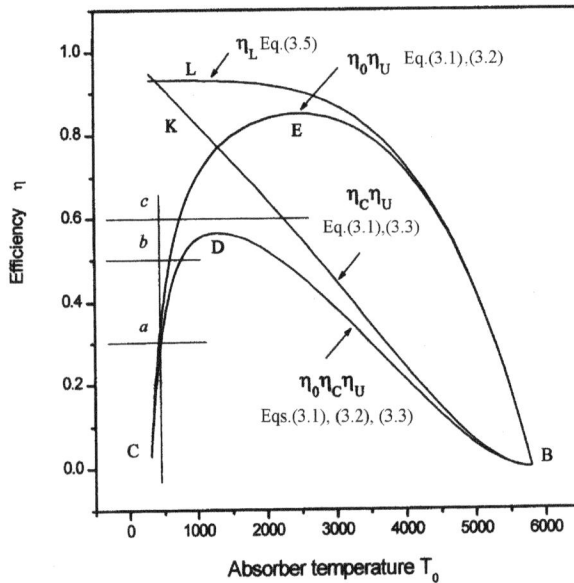

Figure 9: Comparison between efficiencies of reversible and irreversible conversion of solar heat into work. Line "LB" - the limiting efficiency η_L of the reversible process without Carnot cycle according to Equation (3.5). Lines "KB", "CDB" and "CEB" are limiting efficiencies of Carnot cycles as reversible processes occuring in parallel with irreversible solar energy reemission. Points *a,b,c* illustrate the proposed method for calculating the relative contributions of reversible processes in a generally irreversible conversion of solar heat into work [6].

3.2.4. The Diagram of the Solar Cell Efficiency as a Reversible Diagram

The temperature dependence of η_L from equation (3.5) is shown by line "LB" in Fig. (**4**). Let us also take into account that this line is (by definition) a graphic illustration of the sequence of reversible transitions from one state of the system to another only. Then η-coordinates of the points on the lines representing the combinations of efficiencies of the reversible and irreversible processes (lines "KB", "CDB" and "CEB" in Figs. (**3**) and (**4**) are proportional to their share of reversible transitions [6].

For example, let point *a* in Fig. (**4b**) denote the conversion of solar energy along the line "CE" with efficiency of 30%. Let us draw an isotherm through point *a*; the intersections of this line with the line "LB" and the y-axis give us the values $\eta_L = 0.93$ and $T_A = 430$ K. Then, according to the lever principle, the point *a* corresponds to $(\eta_L - 30)100/\eta_L = 68\%$ for irreversible processes and 32% for the reversible ones.

3.3. Antenna States of the Absorber Particles [6]

Let us consider these particle states in a radiant energy absorber. Transitions between them result from the absorption of photons. The states of atomic particles or their groups, as well as energy transitions between them, are called "working" if they take part in performing work. Figs. (**3**) and (**4**) show that solar energy conversions are not always working processes. The states of atomic particles and the energy transitions between them are said to be "antenna" ones if they take part in the absorption and emission of radiant energy without the performance of work. Carnot cycles are examples of working processes, while cycles described below involving the photon re-emission are examples of antenna processes.

It is clear that antenna and working states are equilibrium ones if cycles of the radiant energy conversion are not accompanied by entropy production, *i.e.*, if it takes place along line "LB" in Fig. (**5**) with efficiency η_L from Equation (3.5). Let us suppose their total amount to be 100%. Now assume that the conversion of radiant energy consists of reversible working processes with an efficiency of η_O from the Equation (3.2) and irreversible antenna processes with an efficiency of η_U from the equation (3.1), *i.e.*, that it corresponds to line "CE". In this case, the

η-coordinate of all points divided by η_L is equal to the share of working states, while $1-\eta/\eta_L$ is equal to the share of antenna states.

Fig. (**5**) shows that the number of antenna states (*i.e.*, the number of non-equilibrium states) decreases as efficiency η increases along the line "CE". The isothermal growth of the efficiency η implies that certain antenna states have become equilibrium states and that reversible transitions that do not generate work have appeared. If the temperature dependence of η is determined by experiment, Fig. (**4**) helps to interpret and model the paths along which the solar cell produces work.

Classical works [6] focus on efficiencies of the conversion of solar heat into work by a reversible process (η_L from equation (3.5)) and by Carnot cycles (η_0 and η_C from equations (3.2), (3.3)). However, their authors overlooked one detail: the same isotherm belongs to radiation and matter. We present for the first time a description of the photon re-emission with and without work production. A comparison of its efficiency made it clear that it is impossible to overcome the 30% threshold in solar cell efficiency without reversible reemission.

Let the point *a* on the line "CDB" in Fig. (**5**) denote the conversion of solar energy with an efficiency of 30%. This implies that, when T_A=430 K, the working states account for 32% of all the states of the particles in the absorber, *i.e.*, there are two antenna states for each working one. By the definition of the line "CDB", the working states make up a reversible process. Therefore, the remaining antenna states belong exclusively to irreversible processes. If one were to make reversible processes out of 1/3 of the antenna states, the efficiency η would increase from 30% to 50% (point *b* on the isotherm in Fig. (**5**)). The same effect may be attained along the line "CD" if there is the same number of working and antenna states. However, the absorber would need to have a temperature of 800 K, which is really technically unfeasible.

Let us consider another example. The zone theory proposes mechanisms of converting solar energy into work with an efficiency of η=60%, which cannot be reached by an absorber if it is heated along the lines "CD" and "CE" in Fig. (**5**). Let us denote this value by the point *c* on the isotherm 430 K in Fig. (**5**). It can be reached by an isothermal process only if η=50% of the antenna states constitute a reversible process.

Thus the 30% efficiency threshold in solar energy conversion can be overcome only if solar energy is re-emitted reversibly. In real devices, antenna states can constitute a reversible process only in combination with irreversible ones. Let us therefore recall the definitions of these terms.

The reversibility of a process relates to the second law of thermodynamics and is applicable to any thermodynamic system. Let us recall that the transition of a system from equilibrium state 1 to equilibrium state 2 is called reversible (*i.e.*, bi-directional) if one can return from state 2 to state 1 without making any changes in the surrounding environment, *i.e.*, without compensations. The transition of a system from state 1 to state 2 is said to be irreversible if it is impossible to return from state 2 to state 1 without compensations. Let us now use these notions to describe the equilibrium between a block body and radiation as a continuous and infinite series of quantum electronic transitions.

According to this definition, the re-emission (1231) will alter the frequency distribution of photons in the cavity if states 2 and 3 are energetically distinct. For this reason, such an antenna process is irreversible. The state of radiation will not change if the re-emission (12131) or (12321) can take place. The resulting antenna process will be reversible if the equilibrium state of matter is not violated.

Every quasi-static process is reversible and infinitely slow. Are the continuous series (12131) and (12321) quasi-static processes? For the sake of simplicity, let us approximate the spatial arrangement of particles by a periodic chain of elements that have a length of 2μm and that are separated by 2 Å. The time period during which solar radiation will excite only the first particle in the chain, while subsequent particles are excited only by emitted photons, is equal to 10^{-4} s if the lifetime of the excited state is $\tau=10^{-8}$ s. The time needed for the wave to travel down the chain and excite only the last particle is $\tau=10^{-14}$ s. Therefore, the energy exchange due to re-emission in a chain of particles is an infinitely slow process in comparison with the diffusion of electromagnetic radiation in the chain. Even the multiple repetition of re-emission (121) by one particle during 10^{-6} s is a quasi-static process, for the time taken by such re-emission exceeds by a factor of 100 the lifetime of the excited states.

Thus, the sequences of antenna states of absorber particles in solar cells can take the form of quasi-static processes. Their isolation out of the general process of the conversion of solar energy into work (which is a non-equilibrium process on the whole) does not contradict the laws of thermodynamics.

Let us call the re-emission of solar energy during an antenna process *retranslation*, if the temperature of the radiation remains constant, and *transformation*, if the temperature of the radiation changes. The retranslation of radiation by a black body is, by definition, a reversible process. Transformation can take place both in a reversible and irreversible way. Therefore, the efficiency of the work performed by a solar cell can be improved by increasing the share of antenna as well as working equilibrium states in an absorber.

We propose a special way of overcoming the 30% threshold for solar cell efficiency. It is not directly linked to the improvement of existing types of devices or the search for the optimal physical and chemical properties of the materials. In contrast to previous scholars, we have drawn attention to the fact that the efficiency of a device is also influenced by processes which do not involve the performance of work. We have called them antenna processes.

Antenna processes take place during the conversion of solar heat into work. They can be divided into reversible and irreversible processes theoretically. We have shown that antenna processes can occur in a quasi-static fashion. According to the maximum work principle, an increase in the share of quasi-static antenna processes should lead to the performance of additional work by other processes. The reversibility diagram of the conversion of solar heat into work (Fig. (5)) shows that there are no theoretical prohibitions to attain an efficiency of η_L equal to 93% at room temperature. This diagram will be useful means for the interpretation of the theoretical and experimental investigations of the intermediate band devices and other solar cells [6].

4. CRYSTALLOGRAPHY [6]

4.1. Introduction

Studies of material structure, its optical and thermodynamic properties belong to the scope of several disciplines. For example, crystal lattice is the result of crystal's mathematical modeling, while the regularity of its structure represents the basis of its zone structure, which, in turn, is the model for describing energy states of the electrons. It is clear that after all chemical bonds are broken and crystal is evaporated there will be no crystal lattice and its inherent energy zones. Therefore, studies into relations between structural, thermodynamic and optical properties of material without addressing zone theory of solids are of both theoretical and practical interest. Here we show that such relations can be found by presenting the sequence of ground and excited states of primitive and non-primitive crystal cells as a thermodynamic process. Calculation of frequencies at which maximum useful work can be produced from a crystal cell and radiation is presented on the example of magnesium. The number of such frequencies increases in the row diamond-graphite-soot. On the example of silicon it is suggested that objects can be comparable with ideal radiator, or black body, in terms of their property to absorb radiation energy if they are composed of nanoparticles, *i.e.* clusters with small number of atoms. Requirements are formulated for such objects to be used for maximally efficient conversion of solar radiation into useful work.

4.2. Equilibrium Between Substance and Radiation

In a spectral test with ideal radiator thermodynamic equilibrium between substance and radiation is reached. Therefore, in this case the process of absorption and emission of radiation energy can be viewed as an equilibrium process at all frequencies. In this work it is demonstrated for the first time that the process of radiation energy absorption by a crystal can be an equilibrium process at certain frequencies.

4.3. Equilibrium Between Particles in Crystal

Absorption of radiation energy by crystals is described in terms of zone theory on the basis of thermodynamic equilibrium between material's particles and specially identified crystallographic directions. However, not all cells of a crystal can accommodate geometric forms of zone structure. Even if they can, the cell and the crystal structure may have different symmetries. In the general case, only the primitive Wigner-Zeits cells and elementary Brave cells always accommodate zone theory forms and have identical symmetry with crystal structure. It is important to note that in highly symmetrical crystal structures Brave cells are always non-primitive. Therefore, an ion in a crystal combines the properties of an isolated particle and those of a part of a group. This aspect of ion's behavior in the equilibrium state of a crystal was not considered in earlier works and is described herein below.

4.3.1. Crystal as a Non-Ideal Radiator and Absorber

Let's assume that metal absorbs radiation energy **u** aliquot to internal energy variation $\Delta U(n)$ of one particle with charge n within a cell containing the number of those particles Z, *i.e.*:

$$u = Z \, \Delta U(n) \tag{4.1}$$

where Z is a whole number sequence. If u=hν, Z≠const. and $\Delta U(n)$=const., there will be a set of frequencies, ν*, at which one and the same particle is identically excited, and a different set of frequencies, ν*, will correspond to a different value of $\Delta U(n)$.

4.3.2. Equilibrium Frequencies of a Non-Ideal Radiator

Excitation of crystal at frequencies ν* will initiate different cyclic processes. Now let's limit the sequence of cycles by the duration of experiment, cycle duration - by interval τ, and wave duration – by 1/ν - time interval after which all physical values characterizing the wave repeat themselves, *i.e.* period of oscillation. If 1/ν*=τ/2, the direct and inverse paths of cyclic process coincide. Now let's select those of them which do not affect material's temperature and do not create local heat flows and residual effects in it. These paths can be viewed as infinitely long or slow because the probability of simultaneous excitation of all the material's particles in a spectral test is extremely small. Then, provided that all conditions named above are met, cell excitation at frequencies ν* can be viewed as an equilibrium or quasi-static process.

4.3.3. Useful Work at Equilibrium Frequencies

Absorption at frequency ν^* can be either maximal (as useful work) or less intensive if its reversibility is disrupted by intermediary states. Excitation at frequencies close to ν^* will involve crystal structure defects. Defects are thermodynamically unstable, their absorption is a non-equilibrium process, and is, therefore, less intensive compared to that of cells. As long as these absorption mechanisms occur simultaneously, the absorption factor's dispersion, $\mu = \mu(\nu)$, should have local maximums at frequencies ν^*. Now let's identify them using the following procedure for a metal as a simple crystal chemical system.

4.3.4. Thermodynamic Calculation of Equilibrium Frequencies

For the cells of metal Me_Z, the equilibrium state of free electrons and ions Me^{n+} with charge n at temperature T_s can be presented as:

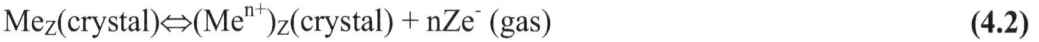

$$Me_Z(crystal) \Leftrightarrow (Me^{n+})_Z(crystal) + nZe^- (gas) \qquad \textbf{(4.2)}$$

Also known for T_s are equilibrium states between crystal and gas at various Z and n. Substituting the difference of internal energies ΔU between any pair of equilibrium states into equation (4.1), it is possible to find ν^* and correlate it with the crystal's value of absorption factor's dispersion $\mu = \mu(\nu)$. If the frequency corresponding to any of the curve maximums coincides with ν^*, the selected pair of equilibrium states with given values of Z and n is realized as a sequence of equilibrium conditions of the crystal's ground and excited states. Let's consider this on the example of Mg.

4.3.5. Mg Film as Non-Ideal Absorber

Mechanisms of magnesium sublimation with various values of Z, n and ΔU were studied previously [6]. By substituting them into equation (4.1), we found u_ν and ν^*. Each value was given a two-digit number, the first digit being Z and the second being n. On the other hand, it is known that absorption factor's dispersion in a Mg thin film is calculated as $\mu = 2u_\nu k/ch$, where k is the imaginary component of the complex dielectric permittivity, c is the speed of light and h is the Planck constant. There are two bands of absorption (Fig. (**1**)). The first (1-11 eV) band involves excitation of the surface plasmon (7.1 eV), while the start of absorption in the second band (49.0 eV) is denoted as $L_{2,3}$ in the band theory.

In Fig. (**1**) the values of u_v (arrows with corresponding numbers) are correlated with values of absorption factor's dispersion, μ. It can be considered that the bands of intensive absorption of radiation energy by magnesium are correlated with the positions of the arrows on Fig. (**10**). For example, the beginning of the 49.0 eV absorption band corresponds to arrow No. 22 which corresponds to the equilibrium states involving Mg^{2+} ions (48.4 eV). Equilibrium states involving atoms (arrows No. 10-50) are correlated with other absorption band. All equilibrium states involving ions Mg^{1+} (arrows No. 11, 21), ions Mg^{2+} (arrows No. 12-52) and Mg^{3+} (arrow No. 13) are preceded by corresponding ionization potentials, I.

There is some discrepancy between the calculated values of u_v and photon energies at the extreme absorption points. For example, for $L_{2,3}$ absorption, this discrepancy is about ~ 1.5 eV. Possible explanation is that ΔU values correspond to massive samples while the values of absorption factor, μ, are corresponding to thin films. Therefore, the difference in optical properties between massive samples and thin films has to be taken into account in this case. Thicknesses of Mg films studied early (100-600Å) were comparable with radiation wavelengths (>100Å) which affected their optical properties, for example, plasmon energy [6].

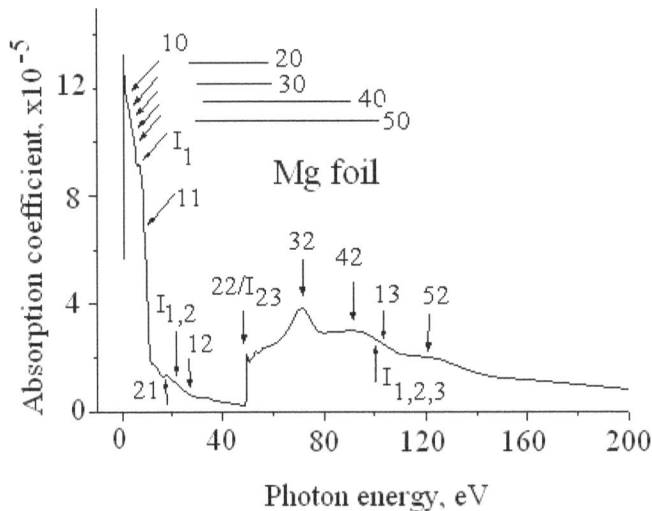

Figure 10: Non-ideal absorber and surface plasmon spectra for Mg nanoparticles [6].

4.3.6. Crystal Symmetry and the Frequency of Equilibrium Absorption

Now let's consider optical and structural properties of the allotropic forms of carbon. Graphite is interesting in that its crystal structure is composed of alternating layers of hexagonal (Z=2) and rhombohedra (Z=3) symmetry. Then, according to (4.1), these two values of Z (n=0) will correspond to increased absorption of photons with energies of 38 eV and 44 eV, respectively. Such absorption was observed in experiments with natural and pyrolytic graphites. Therefore, change in the symmetry of the atom's nearest neighborhood causes a shift in the frequency of equilibrium absorption of radiation energy.

Diamond is of particular interest due to the presence of hexagonal layers of lonsdaylite (Z=2) in its cubic syngony (Z=8) crystal. However, diamond's type of absorption is different from that of graphite. The fact is that in a spectral experiment, diamond and lonsdaylite are thermodynamically unstable form of carbon to which equation (4.1) is inapplicable.

So, it is interesting in that it does not have any crystal structure and its optical properties are close to those of black body, *i.e.* capable of equilibrium absorption of radiation of all frequencies. Soot absorbs stronger than graphite and does not have peculiarities characteristic of graphite at the equilibrium frequencies. Therefore, increase in the proportion of equilibrium frequencies, ν^*, leads to increase in absorption; absorption curve gets smoother, and, in the limit case, should coincide with the absorption curve of black body.

4.3.7. Equilibrium Absorption of Various Chemical Composition Substances

In accordance with (4.1), frequency ν^* is a linear function of ΔU. It has been found that the function $\nu^*=f(\Delta U)$ is similar for Mg and carbon, and the same function describes structural (Z), thermal (ΔU) and optical (ν^*) properties of other metals, semiconductors and dielectrics of various types.

4.3.8. The Future of Light-Absorbing Devices - the Increasing of Equilibrium Absorption

It is important to note that the number of ν^* frequencies of a material can be artificially increased by creating new states by means of physical or chemical treatment. For example, by means of electrochemical etching it is possible to produce quasi-monocrystal silicon with absorption factor by $\sim 10^3$ higher than that of monocrystal silicon at the frequency of $\sim 9,000$ cm^{-1}. It is possible that silicon may absorb radiation with even greater intensity if more equilibrium states by the type and number of particles are created compared to the number of equilibrium states present in crystal or amorphous silicon. Ideally, all cells should be asymmetric, chaotically positioned and different in the geometry and number of particles. A model of such cells can be nanoparticles, *i.e.*, clusters containing small number of atoms (from 2 to a few dozens). Such clusters are of interest today as structures representing an intermediate form of matter with unique properties. In particular, due to minor differences in the dissociation energy, ΔU, such clusters can create a series of adjacently positioned frequencies ν^* in the solid-state system's absorption spectrum; crystal silicon has just one such frequency.

5. CLAPEYRON EQUATION AND ASTROPHYSICAL ASPECT

It is shown that at the condition of non-zero chemical potential of thermal radiation a hidden equilibrium between photon gas and photon condensate is possible. Postulates of thermodynamic will be obeyed if the photon gas is in equilibrium with a condensate and the particles of this condensate are in identical energy state. The nonvariancy of the system "thermal radiation-condensate" means the complete degeneration of the condensate and constant photon and condensate particles numbers ratio at any temperature. Density and energy of condensate particles of well known thermal radiations *e.g.* relic radiation are calculated. We think that the photon condensate hypothesis is useful for the interpretation of dark energy and dark matter which are presently not explained. The completely degenerated Bose-Einstein condensate could be an analog of photon condensate.

The fundamental equation of thermodynamics, the phase rule, the equation of Gibbs-Duham and the Bose statistics are used in studying hidden or supposed equilibrium of thermal radiation. This chapter shows for the first time that thermal radiation possesses some special properties which could make a hidden equilibrium between photon gas and photon condensate possible.

The thermal radiation is an exceptional thermodynamic system because the function from the entropy S and the volume V,

$$dU = TdS - pdV \qquad (5.1)$$

keeps features of the fundamental equation of thermodynamics regardless the photon number is variable [9-12]. Exchange of the parameters in the equation (5.1) on temperature T and pressure p leads to the Gibbs-Duham equation $SdT - Vdp = 0$, which leads to the Clapeyron equation:

$$dp/dT = S/V \qquad (5.2)$$

describing the equilibrium of two phases in the case, when the pressure depends on temperature only. Thermal radiation has dependence,

$$p = \sigma T^4/3 \qquad (5.3)$$

where σ is the Stefan-Boltzmann constant. However, there are no references concerning the two-phase state of radiation.

We assume that equations (5.1)–(5.3) will be obeyed if the photon gas is in equilibrium with a condensate, and the particles of this condensate are in identical energy state. The completely degenerated Bose-Einstein condensate is known as an analogy of such a thermodynamic system.

The phase equilibrium is linked to the chemical potential. A choice of zero and non-zero values of the chemical potential μ is possible in thermodynamics of thermal radiation [12, 13]. It is found out that [13],

$$\mu = 3.602kT \qquad (5.4)$$

The function $\mu_1 = f(T_1)$ of the radiation (phase 1) and the relation $\mu_1 = \mu_2$ for the condensate (phase 2) have solution $T_1 = T_2$ if the dependence $\mu_2 = f(T_2)$, in this case $\mu_2 = 3.602kT_2$, will be assigned to the chemical potential of the condensate. In other cases the condition $T_1 \neq T_2$ is valid preventing equilibrium of the phases 1 and 2.

The pressure is excessive for description of an equilibrium between the photon gas and the condensate due to the relation (5.3). Then the phases rule for the equilibrium "gas-condensate" is n ≤ k+1, where n is a number of phases, k is a number of components. In our case n=2, k=1 and the number of thermodynamic degrees of freedom for the photon system "gas-condensate" is equal (k+1−n) = 0.

The Bose statistics allows to interpret the nonvariance of the photon system "gas-condensate" from the density of photons,

$$N_{ph}/V = \sum_i [g_i / (e^{\beta\varepsilon_i} - 1)] = \pi^{-2} (kT/\hbar c)^3 \int_0^\infty x^2/(e^x-1)dx \qquad (5.5)$$

where g_i is a distribution function for photons on energy ε_i, $x \equiv \varepsilon_i /kT$, β is the Lagrange coefficient equal $1/kT$, k and $\hbar$ are the Boltzmann and the Planck constants, respectively, c is a speed of light in vacuum. The density of photons is found to be as in (5.6) after solving the integral in expression (5.6) [14, 15]:

$$N_{ph}/V = n!\zeta(n+1)(kT/\hbar c)^3/\pi^2 = 2.404 (kT/\hbar c)^3/\pi^2 \qquad (5.6)$$

The particles of condensate are in thermodynamic equilibrium with photons and have the same energy. So, if the condensate is obeying Bose statistics, one must reject the particle summation over energies and apply a variation of the function $y=x^2/(e^x-1)$ in the expression (5.6). It has maximum at $x_m=1.593$ and $y_m=0.6476$. Other values of y are corresponding to other two values of each x. While the system "gas-condensate" is invariant, we consider the expression $x_m=\varepsilon/kT=1.593$ only. Its transformation:

$$\varepsilon_{cond} =1.593kT \qquad (5.7)$$

is the energy of the particle in the condensate. In this case the density of the condensate is,

$$N_{cond.}/V = 2x_m^2 (kT/\hbar c)^3/ [\pi^2(e^{x_m} - 1)] = 1.30 (kT/\hbar c)^3/ \pi^2 \qquad (5.8)$$

Comparing this result with the expression (5.7) we finally receive:

$$N_{ph}/N_{cond}= 1.85$$

According to (5.7), the average density (5.8) of the energy u of the condensate is,

$$u_{cond} = (\varepsilon N)_{cond}/ V = 2.42 \cdot 10^{-16} T^4 = 0.317 \sigma T^4 \qquad (5.9)$$

where σ is the Stefan-Boltzmann constant. In equilibrium u_{cond}/u_{ph} is 0.317 because average density of the photon gas u_{ph} is σT^4. It means that the condensate energy density is $1/0.317 \approx 3.16$ times smaller that the photon gas one at any temperature.

According to (5.8)-(5.9) the invariance of the system "radiation-condensate" means the complete degeneration of the condensate and constant photon and condensate particles numbers ratio at any temperature.

In our hypothesis the photon gas and condensate do not exist without each other. They form a system without phase surfaces. Particles are in equilibrium with matter-thermostat penetrating the internal space of it. Energy of the photon is not dividable. Energy of the condensate particle is also a quantum. Together photon and condensate particle are giving energy completely to the thermostate and transform to states without energy and mass. We call it a proto-quantum. And *vice versa*, proto-quanta get radiant energy from the thermostat and become a photon or a particle of the condensate. Photon appearance and disappearance is the observable part of the dynamic equilibrium "photon gas – condensate". Condensate particles participation in dynamic equilibrium is hidden from us.

It is impossible to prove the existence of condensate particles during a thermodynamic experiment. It is not important now if the photon condensate really exist. Important is the fact that this hypothesis is useful for the interpretation of dark energy and dark matter which are presently not explained.

Table **1** presents density and energy of photon condensate particles. They are calculated according to the expressions (5.7) and (5.9) for characteristic temperatures of well known thermal radiations. For example for the temperature 3000 K when the relic radiation became decoupled from matter and its energy density was higher than the assumed density of dark energy estimated to be 4 GeV/m^3 [16]. At present age the temperature of the relic radiation is 2.725 K. Equilibrium between relic radiation and condensate in the Universe is possible at the condition that their particle velocities are equal. In our opinion photon condensate is the reason for photon distribution in imaginary nothing.

The table shows that the energy density of the relic radiation is comparable with that of the dark energy. It is known [16] that dark energy is homogeneously distributed in the Universe, has a low density and does not noticeable interact by known fundamental types of interaction excluding gravitation. There is no accurate explanation of these phenomena. Vacuum energy or new super weak field, "quintessence", can serve as candidates for dark energy role. Relic condensate has low energy density; it is homogeneously distributed in the Universe and does not exchange energy with matter. It also can be included in the set of candidates for dark energy, more precisely, its component.

The astronomical observations demonstrated gravitational lenses formed not only by the visible mass of galaxies. This gravitational effect is assigned to the dark matter of galaxies, which forms bunches in the galaxies and its properties cannot be explained [16].

A galaxy has 100-200 billions stars. The solar radiation is also thermal radiation. The table demonstrates properties of the solar photon condensate for the temperature 5800 K. One can see that the density of the solar condensate is 10^{10} times greater than that of the relic condensate, and the energy density is $2 \cdot 10^{13}$ times greater. The density of the solar photon gas and condensate decreases with the distance due to unlimited expansion, and the bunch of particles is formed around the Sun. We think that the photon condensate of solar radiations, possessing high density and forming bunches around stars, can be a candidate for the role of that component of dark matter, which forms gravitational lenses.

The properties of photon condensate are calculated for the temperature 10^9 K, as well as for the temperature 10^{15} K at time of nucleosynthesis which was 10^{-9} s after Big Bang. It is interesting that at 300 K a particle of the photon condensate has greater energy than the kinetic energy of the translational movement of the molecule (equal to 0.025 eV).

In summary, one should note that the equilibrium of the photon gas and the photon condensate does not contradict with the laws of thermodynamics. Nevertheless, it is impossible to obtain an evidence of the existence of particles of the condensate in thermodynamic experiment. The conviction of the existence of the photon condensate has been increased after acquaintance with the idea of N. Bogolyubov stating that the particles of the Bose-Einstein gas mainly interact through the condensate and not directly with each other [17, p.109]. The photon condensate with protoquanta can also play the same role for photons. The hypothesis about existence of relic photon condensate with non-zero rest mass of the photon is proposed in [18]. The zero rest mass of the photon does not contradict with the hypothesis about photon condensate described in this paper.

Table 1: Comparison of some properties of the photonic condensate at different temperatures

Temperature	Density $N_{cond.}/V$, Eq. (21)		Energy of particle ε, Eq. (20)		Density of energy u_{cond}, Eq. (22)	
K	particle/m^3	kg/m^3	J	eV	J/m^3	GeV/m^3
2.725	$2.2 \cdot 10^8$	$1.6 \cdot 10^{-31}$	$6.0 \cdot 10^{-23}$	$3.7 \cdot 10^{-4}$	$1.4 \cdot 10^{-14}$	$8.8 \cdot 10^{-5}$
30	$3.0 \cdot 10^{11}$	$2.3 \cdot 10^{-27}$	$6.6 \cdot 10^{-22}$	0.0041	$2.1 \cdot 10^{-10}$	1.3
300	$3.0 \cdot 10^{14}$	$2.3 \cdot 10^{-23}$	$6.6 \cdot 10^{-21}$	0.041	$2.1 \cdot 10^{-6}$	$1.3 \cdot 10^4$
3000	$3.0 \cdot 10^{17}$	$2.3 \cdot 10^{-19}$	$6.6 \cdot 10^{-20}$	0.41	$2.1 \cdot 10^{-2}$	$1.3 \cdot 10^8$
5800	$2.2 \cdot 10^{18}$	$1.0 \cdot 10^{-17}$	$1.3 \cdot 10^{-19}$	0.80	0.29	$5.6 \cdot 10^9$
10^9	$1.1 \cdot 10^{34}$	$2.85 \cdot 10^3$	$2.2 \cdot 10^{-14}$	$1.4 \cdot 10^5$	$2.55 \cdot 10^{20}$	$1.6 \cdot 10^{30}$
10^{15}	$1.1 \cdot 10^{52}$	$2.85 \cdot 10^{27}$	$2.2 \cdot 10^{-8}$	$1.4 \cdot 10^{11}$	$2.55 \cdot 10^{44}$	$1.6 \cdot 10^{54}$

6. THERMODYNAMIC SCALE FOR THE EFFICIENCY OF CHEMICAL ACTION OF SOLAR RADIATION

Radiation energy conversion has a maximum efficiency in natural energy carriers. This efficiency is lower in solar, biological and chemical reactors. The thermodynamic scale of efficiency of chemical action of sun radiation will allow comparing the effectivity of different reactors and estimating their commercial advantages. Such a scale is absent in the well-known thermodynamic descriptions of the solar energy conversion, its storage and transportation to generator of other kinds of energy [18]. In the work the Carnot theorem is a base of the construction of the thermodynamic scale of the efficiency of sun radiation chemical action.

Chemical transformations are related to chemical potentials. In this paper it is shown for the first time that the chemical action of the sun radiation S on the reactant R:

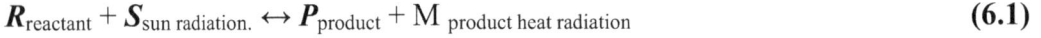

$$R_{\text{reactant}} + S_{\text{sun radiation.}} \leftrightarrow P_{\text{product}} + \text{M}_{\text{product heat radiation}} \qquad (6.1)$$

is so special that the difference of chemical potentials of substances R and P:

$$\mu_R - \mu_P = f(T) \qquad (6.2)$$

becomes a function of their temperature even under idealized reverse process (6.1), if the chemical potential of the sun radiation is accepted to be non-zero.

Actually, there are no obstacles for using the function $f(T)$ in thermodynamic calculations of solar chemical reactions, because previously is shown that the non-zero chemical potential of heat radiation does not contradict with the fundamental equation of the thermodynamics. The sun radiation is a heat radiation.

Let us consider a volume with a black body R, transparent walls, and a thermostat T as an idealized solar chemical reactor. The chemical action of the sun radiation S on the reactant R will be defined by a boundary condition,

$$\mu_R - \mu_P = \mu_m - \mu_S = f(T) \qquad (6.3)$$

where μ_m is a chemical potential of heat radiation of the product P. Then the calculation of the function $f(T)$ is simply reduced to the definition of a difference $(\mu_m - \mu_S)$, while chemical potentials of heat radiations do not depend on chemical composition of the radiator, and the numerical procedure is known [18].

The chemical potential as an intensive parameter of the fundamental equation of the thermodynamics is defined by differentiation of characteristic functions by number of particles N [18].

The intrinsic energy U as a characteristic function of the number of photons $U(V,N) = (2.703 N k_B)^{4/3}/(\sigma V)^{1/3}$ is calculated by the author in [18] by means of compatible solution applied to the known characteristic function $U(S,V) = \sigma V(3S/4 \sigma V)^{4/3}$ [18] and the expression,

$$N = 0.370 \ \sigma T^3 V/k = S/3.602 k_B \tag{6.4}$$

where T, S, V are temperature, entropy and volume of the heat radiation, σ is the Stephan-Boltzmann constant, k_B is the Boltzmann constant. Its derivative,

$$(\partial U/\partial N)_V \equiv \mu = 3.602 k_B T \tag{6.5}$$

introduces a temperature dependence of chemical potential of heat radiation [18]. The function $U(S, V, N)$ as an excerption is not a characteristic one because of the relationship (6.2).

The Sun is a hear radiator of a temperature $T_s = 5800$ K. According to (5), the chemical potential of th sun radiation is $3.602 k_B T = 173.7$ kJ/mole. Then the difference $f(T) = (\mu_m - \mu_S)$ is a function of the measured substance temperature T_m. For example, $f(T) = -165.0$ kJ/mole when $T_m = 298.15$ K, and it is zero when $T_m = T_S$.

According to (6.2), the function $f(T)$ may be presented as the proportional to the dimensionless factor:

$$f(T) = \mu_m - \mu_S = -\mu_S(1 - T_m/T_S). \tag{6.6}$$

according to the Carnot theorem this factor coincides with the efficiency of the Carnot engine $\eta_c(T_m, T_S)$. Then the function:

$$f(T)/\mu_S = -\eta_c(T_m, T_S), \tag{6.7}$$

can characterize an efficiency of the idealized Carnot engine-reactor in the temperature range indicated above.

There is no way to transform heat into work in the heat engine without other changes, *i.e.* without compensation. The heat quantity accepted by the condenser plays this compensation role. If the working substance in the heat engine is a heat radiation with limit temperatures T_m and T_S, then the compensation is presented by the radiation of temperature T_m which is irradiated by the product P at the moment of its formation. Let call this radiation a compensation one in order to make a difference between this radiation and the natural sun radiation.

The efficiency of heat engines with working body consisting of the substance and radiation is considered for the first time in [18]. During the cycle of such an engine-reactor the radiation is cooled from the temperature T_S down to T_m, calling chemical changes in the working substance. The carrier of the energy stored and the compensation radiation are gone out from the engine at the temperature T_m. The efficiency of this heat engine is a base of the thermodynamic scale of the sun radiation chemical action on the working substance.

Assume that the reactant **R** at 298.15 K and the sun radiation **S** of 5800 K temperature are going in the idealized engine-reactor. The product **P** is an output of the engine; this product keeps and transports radiant energy at 298.15 K. The limit working temperatures of the engine are 298.15 and 5800 K. Then, according to relationships (6.2), (6.3), (6.7), the equation:

$$(\mu_m - \mu_S)/\mu_S = -\eta_c(T_m, T_S) \tag{6.8}$$

defines boundary conditions of supporting the chemical reaction steady-state regime at temperature T_m in the idealized Carnot engine-reactor.

According to the Carnot theorem the way in which the working body is supplied with energy, as well as the nature of the working body do not influence the efficiency of the heat engine. The efficiency remains the same under contact heat exchange in the same temperature range. The efficiency of such an idealized engine is:

$$\eta_0(T_m, T_S) = (1 - T_m/T_S). \tag{6.9}$$

Then the relationship of the values η_c and η_0 from (6.8)-(6.9):

$$\zeta = \eta_c/\eta_0 = (\mu_P - \mu_R)/\mu_S/(1 - T_m/T_S). \tag{6.10}$$

is a thermodynamic efficiency ζ of chemical action of the sun radiation the working substance in the idealized engine-reactor.

Let us compare efficiencies ζ of the action of the sun radiation on water in the working cycle of the idealized engine-reactor if the water 298.15 K undergoes the following changes:

$$H_2O_{water} + S_{sun\ rad.} = H_2O_{vapor} + M_{heat\ rad.,.vapor} \tag{6.11}$$

$$H_2O_{water} + S_{sun\ rad.} = H_{2gas} + \tfrac{1}{2}O_{2gas} + M_{heat\ rad.,\ H2} + \tfrac{1}{2}M_{heat\ rad.,O2} \tag{6.12}$$

$$H_2O_{water} + S_{sun\ rad.} = H^+_{gas} + OH^-_{\ gas} + M_{sun\ rad.,\ H^+} + M_{heat\ rad.,\ OH^-} \tag{6.13}$$

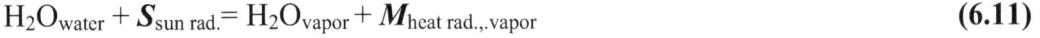

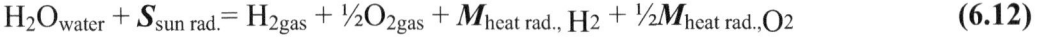

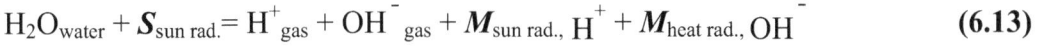

Chemical potentials of substances are equal to the Gibbs energies. In accordance with (6.10),

$$\zeta_{11} = [-228.61-(-237.25)] / 173.7 / 0.95 = 0.052$$

$$\zeta_{12} = \zeta_{11} [0 + \tfrac{1}{2} \cdot 0 - (-228.61)] / 173.7 / 0.95 = 0.052 \times 1.39 = 0.072$$

$$\zeta_{13} = \zeta_{11} [1517.0 - 129.39 - (-228.61)] / 173.7 / 0.95 = 0.052 \times 9.79 = 0.51$$

Thus, in the engine-reactor the reaction (13) may serve as the most effective transformer of the solar energy.

In the real solar chemical reactors the equilibrium of the substance and radiation is not achieved. In this case the moving force of the chemical process in the reactor at the temperature T will be smaller than the difference of Gibbs energies:

$$\Delta G_T = (\mu_P - \mu_R) + (\mu_m - \mu_S) \tag{6.14}$$

For example, water evaporation at 298.15 K under sun radiation is caused by the difference of the Gibbs energies $\Delta G_{298.15} = -228.61-(-237.25) + (-165.0) = -156.4$ kJ/mole. At the standard state (without sun irradiation) $\Delta G^0_{298.15} = (\mu_P - \mu_R) = -228.61-(-237.25) = 8.64$ kJ/mole.

The changes of the Gibbs' energies calculated above have various signs: $\Delta G_{298.15} < 0$ and $\Delta G^0_{298.15} > 0$. It means that the water evaporation at 298.15 K is possible only under the solar process. The efficiency ζ of the solar vapor engine will not exceed $\zeta_{11} = 5.2\%$. There is no commercial advantage because the efficiency of the conventional vapor engines is higher.

However, the efficiency of the solar engine may be higher than that of the vapor one if the condition $\Delta G_{298.15} < 0$ and $\Delta G^0_{298.15} > 0$ is kept. The live cells of plants where photosynthesis takes place is an illustrative example.

Under keeping the condition $\Delta G_{298.15} < 0$ and $\Delta G^0{}_{298.15} < 0$ the radiation heat exchange replaces the chemical action of the sun radiation. If $\Delta G_{298.15} > 0$ and $\Delta G^0{}_{298.15} > 0$, then neither radiation heat exchange, nor chemical conversion of the solar energy call chemical changes in the system under this temperature. The processes (6.12) and (6.13) are examples. Nevertheless, at the temperatures when ΔG becomes negative, the chemical changes will be occurred in the reaction mixture. So, in solar engines-reactors there is a lower limit of the temperature T_m. For example, it had been previously reported that chemical reactors equipped with solr light concentrators have the minimum optimal temperature 1150 K.

The functions $\Delta G(T)$ and $\zeta(T)$ describe various features of the chemical conversion of the solar energy. As an illustration we consider a case when the substance phases *R* and *P* are in thermodynamic equilibrium. For example, the chemical potentials of boiled water and the water vapor are equal. Then their difference ($\mu_P - \mu_R$) and the efficiency of the chemical action of the sun radiation ζ are zero, regardless $\Delta G(T) < 0$. Without irradiation the equation $\Delta G(T) = 0$ determines a condition of the thermodynamic equilibrium, and the function $\zeta(T)$ losses its sense.

The thermodynamic scale of efficiency $\zeta(T)$ of the chemical action of the sun radiation presented here is a necessary criterion for choice of optimal design of the solar engines-reactors. It is simple for application while its values are calculated from the experimentally obtained data of chemical potentials and temperature. Varying the values of chemical potentials and temperature makes it possible to model (with help of expressions (6.10), (6.14)) the properties of the working body, its thermodynamic state and optimal conditions of chemical transformations in solar engines and reactors in order to bring commercial advantages of alternative energy sources.

7. THERMODYNAMIC EFFICIENCY OF PHOTOSYNTHESIS IN PLANT CELLS

Carnot theorem has been used for calculation of the thermodynamic efficiency of the photosynthesis in plants; it is found that the efficiency is 71%.

Sun supplies plants with energy. Only 0.001 of the solar energy reaching the Earth surface is used for photosynthesis [19, 20] producing about 10^{14} kg of green plant mass per year [3]. Photosynthesis is thought to be a low-effective process [4]. The limiting efficiency of green plant is defined to be 5% as a ratio of the absorbed solar energy and energy of photosynthesis products [21, 22].

This chapter of the lecture shows for the first time that the photosynthesis efficiency is significantly higher (71% instead of 5%), and it is calculated as the Carnot efficiency of the solar engine-reactor with radiation and matter as a single working body.

The photosynthesis takes place in the chloroplasts containing enclosed stroma, a concentrated solution of enzymes. Here are occurring the dark reactions of the photosynthesis of glucose and other substances from water and carbon dioxide. The chlorophyll traps the solar photon in photosynthesis membranes. The single membrane forms a disk-like sac, or a thylakoid. It encloses the lumen, the fluid where the light reactions take place. The thylakoids are forming granum [23, 24]. Stacks of grana are immersed into the stroma.

When solar radiation with the temperature T_S is cooled in the thylakoid down to the temperature T_A, the amount of evolved radiant heat is a fraction,

$$\eta_U = 1 - (T_A/T_S)^4 \tag{7.1}$$

of the energy of incident solar radiation [25]. The value η_U is considered here as an efficiency of radiant heat exchange between the black body and solar radiation.

Thylakoids and grana as objects of intensive radiant heat exchange have a higher temperature than the stroma. Assume the lumen in the thylakoid has the temperature T_A=300 K and the stroma, inner and outer membranes of the chloroplast has the temperature T_0=298 K. The solar radiation temperature T_S equals to 5800 K.

The limiting temperatures T_0, T_A in the chloroplast and temperature T_S of solar radiation allow to imagine a heat engine performing work of synthesis, transport and accumulation of substances. In idealized Carnot case solar radiation performs work in thylakoid with efficiency:

$\eta_C = 1 - T_A/T_S = 0.948$ (7.2)

and the matter in the stroma performs work with an efficiency:

$\eta_0 = 1 - T_0/T_A = 0.0067$ (7.3)

The efficiency $\eta_0\eta_C$ of these imagined engines is 0.00635.

The product $\eta_0\eta_C$ equals to the sum $\eta_0 + \eta_C - \eta_{OS}$. Value η_{OS} is the efficiency of Carnot cycle where the isotherm T_S relates to the radiation, and the isotherm T_0 relates to the matter. The values η_{OS} and η_C are practically the same for chosen temperatures, and $\eta_{OS}/\eta_0\eta_C = 150$. It means that the engines where matter and radiation perform work are a single working body has 150 times higher efficiency than the chain of two engines where matter and radiation perform work separately.

It is known [27] that in the idealized Carnot solar engine-reactor solar radiation S produces at the temperature T_A a chemical action on the reagent R:

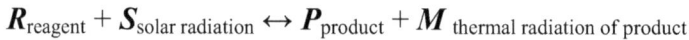

$$R_{reagent} + S_{solar\ radiation} \leftrightarrow P_{product} + M_{thermal\ radiation\ of\ product}$$

with efficiency:

$\zeta = (\mu_P - \mu_R)/\mu_S/(1 - T_A/T_S)$ (7.4)

where μ_P, μ_R are chemical potentials of the substances, μ_S is the chemical potential of solar radiation equal to $3.602 k_B T_S = 173.7$ kJ/mole. The efficiency of using water for alternative fuel synthesis is calculated in [27].

Water is a participant of metabolism. It is produced during the synthesis of adenosine triphosphate (ATP) from the adenosine diphosphate (ADP) and the orthophosphate (P_i):

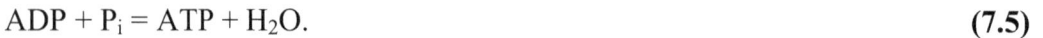

$$ADP + P_i = ATP + H_2O.$$ (7.5)

Water is consumed during the synthesis of the reduced form of the nicotinamide adenine dinucleotide phosphate (NADPH) from its oxidized form ($NADP^+$):

$$2NADP^+ + 2H_2O = 2NADPH + O_2 + 2\ H^+_{thylakoid}$$ (7.6)

and, during the glucose synthesis:

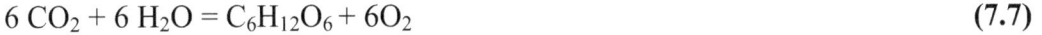

$$6\ CO_2 + 6\ H_2O = C_6H_{12}O_6 + 6O_2 \qquad (7.7)$$

Changes of the Gibbs energies or chemical potentials of substances in the reactions (7.5)-(7.7) are 30.5, 438, and 2850 kJ/mole, respectively [23].

The photosynthesis is an example of joint chemical action of matter and radiation in the cycle of the idealized engine-reactor, when the water molecule undergoes the changes according to the reactions (7.5)-(7.7). According to (7.4), the photosynthesis efficiency ζ_{Ph} in this model is $\zeta_5 \cdot 1/2\zeta_6 \cdot 1/6\zeta_7 = 71\%$.

The efficiency ζ_{Ph} is smaller than the Landsberg limiting efficiency:

$$\eta_L = \eta_U - 4(T_0/T_S - T_0\ T_A^3/T_S^4)/3 \qquad (7.8)$$

known in the solar cell theory [25] as the efficiency of the joint chemical action of the radiation and matter per cycle. ζ_{Ph} and the temperature dependence η_L are shown in Fig. (**1**) by the point F and the curve "LB", respectively. They are compared with the temperature dependence of efficiencies $\eta_0\eta_C\eta_U$ (curve "CB") and $\eta_{OS}\eta_U$ (curve "KB"). Value η_U is close to unity because $(T_A/T_S)^4 \sim 10^{-5}$.

We draw in Fig. (**11**) an isotherm "tt'" of η values for $T_A=300$ K. It is found that $\eta_{OS} = 94.8\%$ at the interception point K, $\eta_L=93.2\%$ at the point L, and $\eta_0\eta_C\eta_U=0.635\%$ at the point C. The following question arises: which processes give the chloroplasts energy for overcoming the point C and achieving an efficiency $\zeta_{Ph} = 71\%$ at the point F?

First of all one should note that the conversion of solar energy into heat in grana has an efficiency η_g smaller than η_U of the radiant heat exchange for black bodies. From (7.8) follows that the efficiency ζ_{Ph} cannot reach the value η_L due to necessary condition $\eta_g < \eta_U$.

Besides in the thylakoid membrane the photon reemissions take place without heat dissipation [23, 24]. The efficiency area between the curves "LB" and "CB" relates to photon re-emission or antenna processes. They can be reversible and irreversible. The efficiencies of reversible and irreversible processes are different.

Then the point F in the isotherm tt' is the efficiency of engine with the reversible and irreversible antenna cycles.

The antenna process performs the solar photon energy transfer into reaction centre of the photosystem. Their illustration is given in [23, 24]. Every photosystem fixes from 250 to 400 pigments around the reaction center [28]. In our opinion, a single pigment performs reversible or irreversible antenna cycles. The antenna cycles form antenna process. How many pigments make the reversible process in the photosynthetic antenna complex?

One can calculate the fraction of pigments performing the reversible antenna process if the line "LC" in Fig. (**11**) is supposed to have the value equal unity. In this case the point F corresponds to a value $x=\zeta/(\eta_L-\eta_0\eta_C\eta_U)=0.767$. This means that 76.7% of pigments make the reversible antenna process. 23.3% of remaining pigments make an irreversible energy transfer between the pigments to the reaction centers.

The radiant excitation of electron in photo-system occurs as follows:

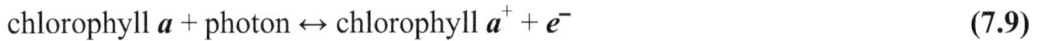

$$\text{chlorophyll } \boldsymbol{a} + \text{photon} \leftrightarrow \text{chlorophyll } \boldsymbol{a}^+ + \boldsymbol{e}^- \tag{7.9}$$

The analogous photon absorption takes place also in the chlorophylls **b**, **c**, **d**, various carotenes and xanthophylls contained in different photo-systems. The excitation of an electron in the photo-systems P680 and P700 are used here as illustrations of the reversible and irreversible antenna processes.

Schemes of working and antenna cycles are shown in Fig. (**2**). Working pigment (a) is excited by the photon in the transition 1→3. Transition 3→2 corresponds to the heat compensation in the chloroplast as engine-reactor. The evolved energy during the transition 2→1 is converted into the work of electron transfer or ATP and NADPH synthesis.

When the antenna process passes beside the reaction centre, the photo-systems make the reversible reemission. Fig. (**12**) presents an interpretation of absorption and emission of photons in antenna cycles. The re-emission 232 shows a radiant heat exchange. The re-emission 121 and 131 takes place according to (7.9). Examples are the pigments in chromoplasts.

According to the thermodynamic postulate, the efficiency of reversible process is limited. In our opinion, just the antenna processes in the pigment molecules of the thylakoid membrane allow the photo-systems to overcome the forbidden line "CB" (for a heat engine efficiency) in Fig. **(1)** and to achieve the efficiency ζ_{Ph} = 71% in the light and dark photosynthesis reactions.

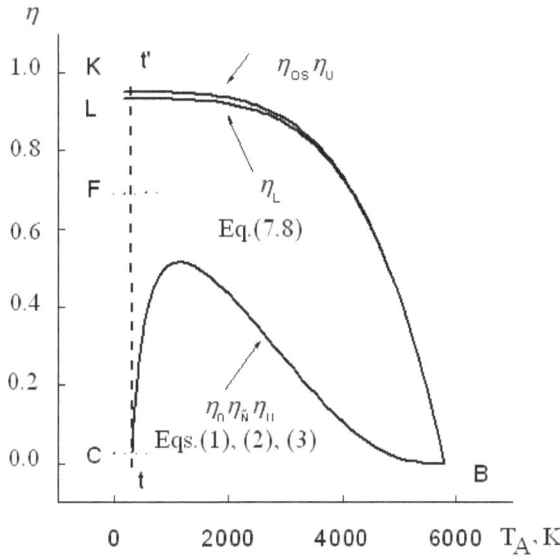

Figure 11: Curve "CB" is the efficiency of the two Carnot engines [30]. The curve "LB" is the efficiency of the reversible heat engine in which solar radiation performs work in combination with a substance [25]. The curve "KB" is the efficiency of the Carnot solar engine-reactor [26], multiplied by the efficiency η_U of the heat exchange between black bodies. The isotherm tt' corresponds to the temperature 300 K. The calculated photosynthesis efficiency is presented by the point F in the isotherm.

There are no difficulties in taking into account in (4) the features of the photosynthesis in different cells. The efficiency of glycolyse, Calvin and Krebs cycles in various living structures may be calculated by the substitution of solar radiation chemical potential in the expression (7.4) by the change of chemical potentials of substances in the chemical reaction.

The cell is considered in biology as a biochemical engine. A definition of biochemical engine is not found. Chemistry and physics know attempts to present the plant photosynthesis as a working cycle of a solar heat engine [29]. The physical action of solar radiation on the matter of non-living systems during working cycle of the heat engine is described in [30, 31]. One can hope that the theoretical investigation of antenna states of chloroplasts in the plant cell made in this work will open new ways for improving new technologies of solar cells and synthesis of alternative energy sources from the plant stuff.

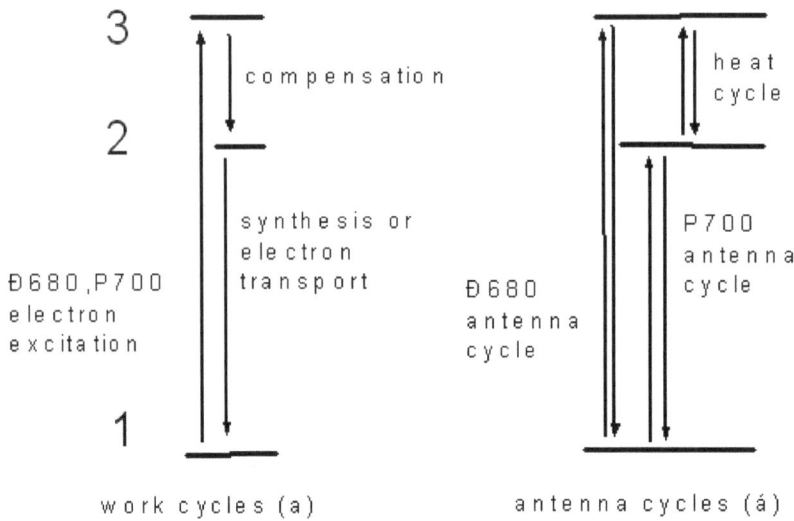

Figure 12: The interpretation of energy transitions in the work (a) and antenna (b) cycles. Level 1 shows the ground states, levels 2, 3 present excited states of pigment molecules.

ACKNOWLEDGEMENTS

Declared none.

CONFLICT OF INTEREST

The author(s) confirm that this chapter content has no conflict of interest.

REFERENCES

[1]   Goetzberger, A.: Knobloch J.; Voss, B. *Crystalline Silicon Solar Cells*; John Wiley & Sons: Chichester, **1998**.

[2]   Lin, Y.-Y.; Chen, C.-W.; Chu, T.-H.; Su, W.-F.; Lin, C.-C.; Ku, C.-H.; Wu, J.-J.; Cheng-Chen, C.-H. Nanostructured metal oxide/conjugated polymer hybrid solar cells by low temperature solution processes. *J. Mater. Chem.*, **2007**, *17*, 4571–4576.

[3]   Chen, G.-Y.; Lee, M.-W.; Wang, G.-J. Fabrication of Dye-Sensitized Solar Cells with a 3D Nanostructured Electrode. *International Journal of Photoenergy*, **2010**, *Volume 2010*, Article ID 585621, 7 pages.

[4]   Yang, K.; El-Emawy, M.A.; i Gu, T.-Y.; Andreas Stintz, A.; Lester, L.F. GaAs Based InAs/InGaAs Quantum Dots-in-a-Well Solar Cells and Their Concentration Applications In: Proceedings of Mater. Res. Soc. Symp., **2010**, *Vol. 1211*, 1211-R03-02-1-6.

[5]   Morisawa, R.; Shirakura, A.; Du, C.-C.; Huang, J.-R.; Liang, M.-W.; Wu,D.C.; Suzuki, T. Structure and optical characterization of nanocrystalline silicon thin films for solar cells. In: Proceedings of Mater. Res. Soc. Symp., **2010**, *Vol. 1211*, 1211-R03-12-1-6.

[6]   Laptev, V.I.; Khlyap, H. In: *Solar Cell Research Progress*. Josef A. Carson, Ed.; Nova Science Publishers: New York, **2008**; pp.181-204.

[7]   Thøgersen, A.; Bonsak, J.; Mayandi, J.; Marstein, E.S.; Mahalingam Umadevi, M. Characterization of Ag nanocrystals for use in solar cell applications. In: Proceedings of Mater. Res. Soc. Symp., **2010**, *Vol. 1211*, 1211-R11-37-1-6.

[8]   Grassoa, C.; Nanu, M.; Goossens, A.; Marc Burgelman, M. Electron transport in $CuInS_2$-based nanostructured solar cells. *Thin Solid Films*, **2005**, *480–481*, 87-91.

[9]   Muenster, A. *Classical Thermodynamics*; Wiley: New York, **1970**.

[10]   Bazarov, I.P. *Thermodynamics*; Oxford: Pergamon, **1964**.

[11]   Kondepudi, D.; Prigogine, I. *Modern Thermodynamics: From Heat Engines to Dissipative Structure*; Wiley: New York, **1998**.

[12]   Laptev, V. I. Chemical Potential and Thermodynamic Functions of Thermal Radiation. *Russion Journal of Physical Chemistry A*, **2010**, *84 (2)*, 158-162.

[13]   Couture, L.; Zitoun, R. *Statistical thermodynamics and properties of matter*; Gordon and Breach: Amsterdam, **2000**.

[14]   Mazenko, G.F. *Equilibrium statistical mechanics*; Wiley: New York, **2000**.

[15]   Rubakov, V.A. *Unsolved problems of cosmology*; **2009**, http://ppc.inr.ac.ru/data/fizfak-3oct-09.pdf

[16]   Bogolyubov, N. *Statistical mechanics*; Complete works in 12 volume set. Nauka Publishers: Moscow, **2006**; vol. 6.

[17]   Kuzmin, V.A.; Shaposhnikov, M.E. Condensation of photons in the hot universe and longitudinal relict radiation. *JETP Lett.,* **1978**, *27*, 628-631.

[18]   Laptev, V.I. Thermodynamic Scale of the Efficiency of Chemical Action of Solar Radiation. *Doklady Physical Chemistry,* **2009***, 429, 243–245*

[19]   Lehninger, A.L.; Nelson, D.L.; Cox, M.M. *Principles of biochemistry*; Freeman: New York, **2008**.

[20]   Pechurkin, N.S. *Energiya i zhizn'*; Nauka Publishers: Novosibirsk, **1988**.

[21]   Odum, E.P. *Basic ecology*; CBS College Publishing: New York, **1983**.

[22]   Ivanov, K.P. Energiya i zhizn'. *Uspekhi sovremennoi biologii* (Russia), **2008**, 128(6), 606-619.

[23]   Voet, D.J.; Voet, J.G.; Pratt, C.W. *Principles of biochemistry*; John Wiley & Sons, Inc. : New York, **2008**.

[24]   Berg, J.M.; Tymoczko, J.L.; Stryer, L. *Biochemistry*; Freeman: New York, **2006**.

[25]   Wuerfel, P. *Physics of Solar Cells*; WILEY-VCH Verlag GmbH & Co. KGaA: Weinheim, **2005**.

[26]   Laptev, V.I. The Special Features of Heat Conversion into Work in Solar Cell Energy Reemission. *Russion Journal of Physical Chemistry*, **2006**, 80(7), 1011-1015.

[27]   Laptev, V.I. Thermodynamic Efficiency of the Photosynthesis in Plant Cell. *Doklady Physical Chemistry*, **2010**, *430*, 36-38.

[28]   Raven, P.H.; Evert, R.F.; Eichhorn, S.E. *Biology of Plants*; Freeman: New York, **2003**.

[29]   Landsberg, P.T. A note on the thermodynamics of energy conversion in plants. *Photochemistry and Photobiology*, **1977**, 26(3), 313-314.

[30]   Laptev, V.I. Conversion of solar heat into work: A supplement to the actual thermodynamic description. *J. Appl. Phys.*, **2005**, *98*, 124905-1-5.

[31]   Laptev, V.I. In: *Solar Cell Research Progress*; Joseph A. Carson, Ed.; Nova Science Publishers, Inc.: New York, **2008**; pp. 131-179.

Send Orders of Reprints at reprints@benthamscience.net
From Semiclassical Semiconductors to Novel Spintronic Devices, 2013, 97-136   **97**

CHAPTER 3

Modeling of Thin Film-Based Semiconductor Active Elements

Halyna Khlyap [*]

Distelstr.12, D-67657 Kaiserslautern, Germany

Abstract: This lecture comprises basic fundamentals of the semiconductor devices' modeling and illustrates the most interesting practical applications for novel spintronics structures.

Keywords: Semiconductor thin film, basic equations, current-voltage characteristic, thin film-based field effect transistor.

1. THE CLASSICAL SEMICONDUCTOR EQUATIONS

The fundamental semiconductor equations [1, 2] which form the basis of the majority of physical device models, may be obtained from an approximate solution for the first two moments of the Boltzmann transport equation [1]:

$$(\partial v/\partial t) + v\nabla v + (qE/m^*) + (1/m^*n)\nabla(nk_BT_e) = (\partial v/\partial t)_{coll} \qquad (1.1)$$

Here the collision term is defined as [1]:

$$\left(\frac{\partial \overrightarrow{v}}{\partial t}\right)_{coll} = -\frac{v - v_0}{\tau},$$

where m^* is the effective mass, k_B is the Boltzmann constant, T_e is the electron temperature, **E** is the electric field, τ is the relaxation time, v_0 is a spherically symmetric solution related to the group velocity **v** [1].

The equation of the momentum conservation (1.1) is further simplified in the drift-diffusion approximation by assuming that the electron temperature gradient ∇T_e is negligible and that the term $v\nabla v$ is small compared with other terms in the equation. Supposing that there is at least an order of magnitude difference between the device and circuit responses: thus, it implies that a quasi-steady-state

---

*Address correspondence to Halyna Khlyap: Distelstr.11, D-67657 Kaiserslautern, Germany; Tel: +49 631 414 4865; Email: gkhlyap17@yandex.ru

model is adequate for most purposes and hence it is convenient to assume that $(\partial \mathbf{v}/\partial \mathbf{t}) = 0$. Finally it is assumed that the electron temperature is equal to the lattice temperature T. Under these circumstances, the equation (1.1) for electrons reduces to the more familiar form [1]:

$$V_n = -\mu_n E -(D_n/n)\nabla n \tag{1.2}$$

where the electrons are treated as negatively charged particles of charge $-q$ Coulombs and the electron mobility μ_n is defined as [1]:

$$\mu_n = q\tau_n/m^* \tag{1.3}$$

The diffusion coefficient for electrons is defined as [1]:

$$D_n = k_B T\mu_n/q \tag{1.4}$$

Here T is the lattice temperature. The.lattice temperature is generally assumed to be constant in classical models. This representation of the carrier transport is often referred to as the 'drift-diffusion' model [1].

$$\partial n/\partial t = (1/q)\nabla \mathbf{J}_n + G \; \textit{for electrons} \tag{1.5}$$

$$\partial p/\partial t = (-1/q)\nabla \mathbf{J}_p - G \; \textit{for holes} \tag{1.6}$$

Here $\mathbf{J}_n$ and $\mathbf{J}_p$ are the electron and hole current densities, respectively, and G is the generation-recombination rate [1].

Current densities are written as follows:

$$J_n = qn\mu_n E + qD_n\nabla n \; \textit{for electrons} \tag{1.7}$$

$$J_p = qp\mu_p E - qD_p\nabla p \; \textit{for holes} \tag{1.8}$$

In these expressions μ_n and μ_p are the electron and hole mobilities, D_n and D_p are the carrier diffusion coefficients [1]. In general terms, the classical carrier transport models require the self-consistent solution of these equations [1].

Equations (1.7) and (1.8) are often referred to as semiconductor mobility equations. The electric field **E** is given by the expression (1.9) [1]:

$$\mathbf{E} = -\nabla\psi \tag{1.9}$$

The potential distribution and electric field are related to the charge in device by the Poisson's equation (1.10) [1]:

$$-\nabla^2\psi = (q/\varepsilon_0\varepsilon_r)(N_D - n + p - N_A) \tag{1.10}$$

Here ψ is the electrostatic potential, q is the charge of the electron, $\varepsilon_0\varepsilon_r$ is a permittivity, N_D is the donor doping density, N_A is the acceptor doping density, n is the electron concentration, and p is the hole concentration [1].

The diffusion coefficients D_n and D_p are usually defined using the Einstein relationships:

$$D_n = \mu_n k_B T/q \tag{1.11}$$

$$D_p = \mu_p k_B T/q \tag{1.12}$$

Here k_B is the Boltzmann constant, and T is the lattice temperature. A common variation on these equations is used to replace the carrier mobility-electric field products by the drift velocity terms (due to the applied electric field) [1],

$$\mathbf{v}_{nd} = -\mu_n\mathbf{E} \tag{1.13}$$

$$\mathbf{v}_{pd} = \mu_n\mathbf{E} \tag{1.14}$$

where v_{nd} and v_{pd} are the electron and hole drift velocities, respectively [1]. We have to note that the electrons are negatively charged particles (with charge $-q$ Coulombs). This velocity term should not be confused with the total velocities v_n and v_p due to the field and diffusion components defined in previous equations [1].

The semiconductor equations are often normalized to simplify the expressions, leading to a more straightforward algorithm saving on computer time [1]. The normalized semiconductor equations are usually expressed in the following form [1]:

$$\partial n/\partial t = \nabla\mathbf{J}_n + G \tag{1.15}$$

$$\partial p / \partial t = -\nabla \mathbf{J}_p - G \tag{1.16}$$

$$\mathbf{J}_n = (\gamma_n)^{-1}(-n\nabla\psi + \nabla n) \tag{1.17}$$

$$\mathbf{J}_p = (\gamma_p)^{-1}(-p\nabla\psi - \nabla p) \tag{1.18}$$

$$\nabla^2\psi = n - p - N \tag{1.19}$$

where $(\gamma_n)^{-1}$ and $(\gamma_p)^{-1}$ are the normalized carrier nobilities [1]. The expressions for current densities $\mathbf{J}_n$ and $\mathbf{J}_p$ assume the Einstein relationship for the diffusion coefficients [1].

The quasi-Fermi level approach is one of the most useful algorithms for simulations. The quasi-Fermi levels for electrons ψ_n and holes ψ_p are as follows [1]:

$$\psi_n = \psi - \ln(n) \tag{1.20}$$

$$\psi_p = \psi + \ln(p) \tag{1.21}$$

To simplify the solution procedure, the equations (1.15), (1.16), (1.19) may be reduced to three coupled elliptic partial differential equations [1] by introducing the new variables φ_n and φ_p, where:

$$\varphi_n = \exp(-\psi_n) \tag{1.22}$$

$$\varphi_p = \exp(\psi_p) \tag{1.23}$$

Boltzmann' statistics have been used to calculate the carrier concentrations n and p and to the description of the potential ψ, for nondegenerate semiconductors with parabolic band structures at thermal equilibrium [1]:

$$n = n_i\exp(q\psi/k_BT) \tag{1.24}$$

$$p = n_i\exp(-q\psi/k_BT) \tag{1.25}$$

Here n_i is the intrinsic carrier concentration at thermal equilibrium and it is given by [1]:

$$n_i = (np)^{1/2} = (N_cN_v)^{1/2}\exp(-E_g/2k_BT) \qquad (1.26)$$

where N_c and N_v are the effective densities of states in the conduction and valence bands, respectively, E_g is the energy band gap of the semiconductor [1]. A non-linear Poisson equation is then written as follows [1]:

$$-\nabla^2\psi = (q/\varepsilon_0\varepsilon_r)[-n_i\exp(q\psi/k_BT) + n_i\exp(-q\psi/k_BT) + N_D - N_A] \qquad (1.27)$$

Semiconductor device models described by these equations assume that the carrier velocities are instantaneous functions of the electric field and that the mobility and diffusion coefficients are functions of the electric field alone [1]. In practice the carriers do not respond instantly to changes in the electric field, and we have take into account the tensor character of the main parameters of semiconductor devices [1].

2. SEMICONDUCTOR EQUATIONS FOR HETEROJUNCTIONS

As is known, the micro(nano)scaled semiconductor heterostructures are the most important building blocks of the novel electronics. Numerical simulation of these devices has special features. Now we consider sets of the equations numerically describing the semi-classical heterostructures.

GaAs-based heterojunctins are traditional objects for which the mathematical models had been successfully developed. One of the main starting points for deriving the set of equations is the construction of the energy band diagram. The numerical methods and their applications for real semiconductor heterostructures are presented in detail in [2]. We would like to draw reader's attention to the energy band diagram of p-GaAs/n-AlGaAs abrupt heterojunction (Fig. (**1**)) [1]. Band-bending occurs because of charge redistribution caused by the requirement to maintain continuity of the Fermi level [1]. As the band gaps are different for the two materials, a discontinuity occurs at the interface producing a 'spike' and 'notch' in the conductiomn band [1].

To simulate heterojunctions we have to take into account the following principal parameters; the band gap, electron affinity, permittivity, mobility and diffusion coefficients. These parameters have a positional dependence in heterostructures

[1]. Furthermore, there is a built-in field due to changes in the band gap and significant interface recombination in the vicinity of the 'notch' (Fig. (**1**)) [1].

In a heterojunction device the relative permittivity ε_r varies according to the position in the device and the composition of the material [1].Thus, we have to modify the Poisson's equation to take into account the inhomogeneities of the heterostructure components. This equation can be rewritten as follows [1]:

$$\nabla(\varepsilon_0\varepsilon_r\mathbf{E}) = \rho \tag{2.1}$$

then [1]:

$$E\nabla\varepsilon_0\varepsilon_r + \varepsilon_0\varepsilon_r\nabla E = \rho \tag{2.2}$$

To simplify the analysis we consider the one-dimensional case where,

$$E_x(d\varepsilon_0\varepsilon_r/dx) + \varepsilon_0\varepsilon_r(dE_x/dx) = \rho(x) \tag{2.3}$$

substituting for the electric field in terms of the potential ψ [1],

$$d^2\psi/dx^2 = (q/\varepsilon_0\varepsilon_r)(N_D - n + p - N_A) - (1/\varepsilon_0\varepsilon_r)(d\varepsilon_0\varepsilon_r/dx)(d\psi dx) \tag{2.4}$$

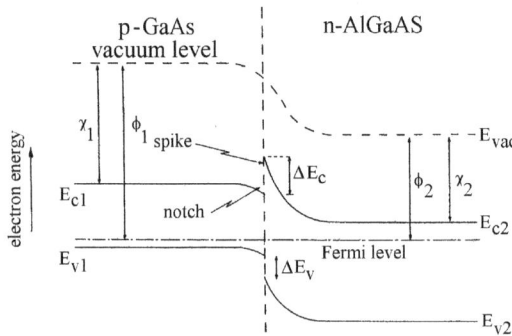

Figure 1: Energy band diagram of an abrupt anisotype heterojunction p-GaAs/n-AlGaAS at thermal equilibrium [1].

The equations for current densities are also to be modified; we have to take into account the energy band structure in the inhomogeneous active element. Assuming the thermal equilibrium for a non-degenerate semiconductor, the electron and hole current densities can be expressed in terms of the gradients of the quasi-Fermi levels E_{fn} and E_{fp} [1]:

$$\mathbf{J}_n = -n\mu_n \nabla E_{fn} \tag{2.5}$$

$$\mathbf{J}_p = -p\mu_p \nabla E_{fp} \tag{2.6}$$

The electron and hole Fermi levels are given by [1]:

$$E_{fn} = q\psi - (\chi - \chi_r) + k_B T \exp(n/N_c) \tag{2.7}$$

$$E_{fp} = -q\psi - (\chi - \chi_r) + E_g - k_B T \exp(p/N_v) \tag{2.8}$$

where χ is the electron affinity, χ_r is the reference level, E_g is the band gap, N_c and N_v are the conduction and valence band densities of states [1]. Using previous equations we obtain for the current densities [1]:

$$\mathbf{J}_n = \mu_n n[q\mathbf{E} - \nabla\chi + (k_B T/n)\nabla n - (k_B T/N_c)\nabla N_c] \tag{2.9}$$

$$\mathbf{J}_p = \mu_p p[q\mathbf{E} - \nabla\chi - \nabla E_g + (k_B T/p)\nabla p - (k_B T/N_v)\nabla N_v] \tag{2.10}$$

Further we describe how the quantum mechanics is applied for numerical simulation of semiconductor devices.

3. APPLICATION OF QUANTUM MECHANICS TO SEMICONDUCTOR DEVICE MODELING

The basic equations used here are the Poisson equation giving the potential distribution $V(x, y, z)$ [1]:

$$\nabla^2 V = (-q/\varepsilon_0\varepsilon_r)(N_D - n) \tag{3.1}$$

and the time independent Schrödinger equation producing the wave function and subband energies [1]:

$$(\hbar^2/2m)\nabla^2\psi + \psi_n(E_n + qV) = 0 \tag{3.2}$$

Here ψ_n is the wave function corresponding to subband n and E_n is the energy of the bottom of the subband *n* [1]. The electron distribution n(z) is then obtained from [1]:

$$n(z) = \sum_n N_n |\psi_n(z)|^2 \tag{3.3}$$

where N_n is the occupation of subband n by the given Fermi distribution [1].

A deterministic model fo the self-consistent treatment of electron propagation in small scale semiconductor devices neglects carriet scattering and assumes simple band structure [1]. The model obtains a self-consistent solution of the electron density and electrostatic potential by solving the wave equation in conjunction with Poisson equation [1]. A one-dimensional device is modeled, with two ohmic contacts in thermal equilibrium [1]; the contacts launch electrons into the device with a range of wave vectors k, and the electron wave function in the device is given by [1]:

$$\Psi(\mathbf{r}) = \psi(x)\exp(ik_t r_t) \tag{3.4}$$

which is obtained by solving the following equation [1]:

$$d[(\gamma(x)^{-1})(d\psi(x)/dx)]/dx = (2m^*/\hbar^2)[E_p + E_t(1 - E_t(1 - \gamma(x)^{-1}) - E_c(x)]\psi(x) = 0 \tag{3.5}$$

where $\psi(x)$ is the envelope function, x_c is the contact position, $\gamma(x)$ describes the spatial variation of the effective mass with respect to that in the contact $m^*(x_c)$ where [1]:

$$\gamma(x) = m^*(x)/m^*(x_c) \tag{3.6}$$

The transverse energy E_t and longitudional energy E_p are given by [1]:

$$E_t = \hbar^2(k_t)^2/2m^*(x_c), \quad E_p = \hbar^2(k_x)^2/2m^*(x_c) \tag{3.7}$$

If it is assumed that the effective mass $m^*(x)$ is position independent, then $\psi(x)$ will be independent of k_t and the electron density for electrons impinging from the left will be described by the expression [1]:

$$n(x)^{1-r} = \frac{1}{2\pi}\int_0^\infty dk_x |\psi_{k_x}^*(x)|^2 \sigma(k_x) \tag{3.8}$$

where $\sigma(k_x)$ is given by [1]:

$$\sigma(k_x)=(m^*(x_c)k_BT)/(\pi\hbar^2)\ln[1+\exp\{(E_{FL}-E_{CL}-E_p)/k_BT\}] \tag{3.9}$$

here E_{FL} is the Fermi level at the left contact and k_BT is the thermal energy [1]. The total electron density is obtained as a function of position by adding the contribution of the left contact to that due to the right contact [1]. The integral in the equation (3.8) is evaluated by summing the contributions to n(x) for each k_x [1].

Detailed quantum mechanical models have been developed to study amorphous materials [1]. If the wave function is expanded as,

$$\psi(\mathbf{r}, t) = (\Omega^{-1}) \sum_k a_k(t)\exp(i\mathbf{kr}) \tag{3.10}$$

where Ω is the volume of the system [1]. The amplitudes satisfy the equations (3.11), (3.12) [1]:

$$i\hbar\partial a_k/\partial t-(\hbar^2k^2/2m)a_k - \sum_k V(\mathbf{k'}-\mathbf{k})a_{k'} = 0 \tag{3.11}$$

$$V(\mathbf{k}-\mathbf{k'}) = (\Omega^{-1}) \int d\mathbf{r}V(\mathbf{r})\exp(i(\mathbf{k'}-\mathbf{k})\mathbf{r}) \tag{3.12}$$

Here $V(\mathbf{r})$ is the local potential [1]. The density of states is obtained by setting the amplitudes at t = 0 to have the form $\exp(i\varphi_k)$, where φ_k is a random phase [1]. The coarse-grained density of states as a function of energy E and averaged over all sets of random phases at T>>t is given by [1]:

$$g(E) = \frac{1}{\pi\hbar}\mathrm{Re}\overline{\int_0^T \exp\left(-\frac{t}{\tau}\right)\exp\left(\frac{iEt}{\hbar}\right)\sum_k \exp(-i\phi_k)a_k(t)dt} \tag{3.13}$$

In the next chapters we present basic models for building parts of the novel microelectronics: high electron mobility transistor (HEMT), field-effect transistor (FET), heterojunction bipolar transistor (HBT) as well as some aspects of carrier transport in disordered bulk and thin film materials.

4. HIGH ELECTRON MOBILITY TRANSISTOR

High electron mobility transistors (HEMT) are belonged to the family of field effect transistors (FET) based on single-junction heterostructures [3, 4].

Historically, the materials from the group A3B5 were pioneer components of HEMTs.

The principal structure of the HEMT is shown in Fig. (2) [3]. The multilayer system is constructed on the semi-insulating GaAs substrate: a thin GaAs layer is deposited as a first one, then an undoped AlGaAs film, then a doped AlGaAs layer; another thin GaAs film finalizes this system [3]. Deposited aluminum forms a Schottky barrier serving as a gate, and provides two Ohmic contacts working as source and drain [3].

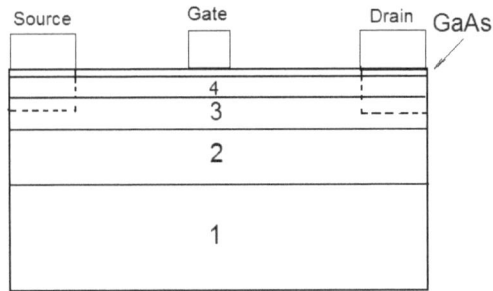

Figure 2: The structure of a HEMT: 1 – semi-insulating GaAs substrate, 2 – p-GaAs layer of 1 μm thickness, 3 – 60 Å undoped $Al_{0.3}Ga_{0.7}As$ film, 4 – 500 Å $Al_{0.3}Ga_{0.7}As$ layer with carrier concentration $n = 7\times10^{17}$ cm^{-3} [3].

Here the quantum well is formed by the GaAs layer neat the interface with the $Al_{0.3}Ga_{0.7}As$. Carriers are supplied by the doped $Al_{0.3}Ga_{0.7}As$ layer [3].

The voltage applied between the gate and the source governs the current flowing from the source and the drain. This applied bias causes the charges redistribution leading to the alternation of the carriers density and, therefore, the current through the channel formed by the quantum well in the GaAs layer near the heterointerface [3]. These factors are to be taken into account to derive the expression for the current-voltage characteristic of the HEMT.

4.1. Potential Distribution and Accumulated Charge Density [3]

Assume the negative voltage V_g is applied between the gate and the GaAs layer (Fig. (3)) [3].

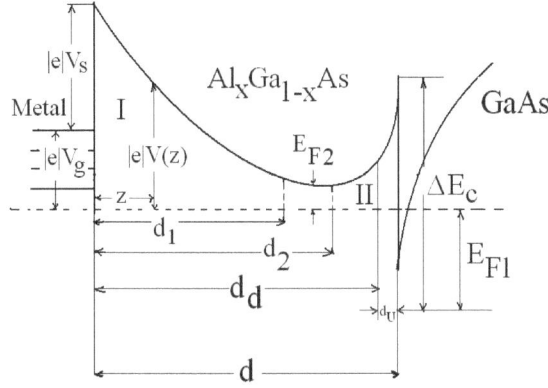

Figure 3: Potential distribution in a HEMT: ΔE_c is a conduction band offset between GaAs and $Al_xGa_{1-x}As$, E_{F1} and E_{F2} are the Fermi levels in $Al_xGa_{1-x}As$ and GaAs layers, respectively, a voltage V_g is assumed to be applied to the gate [3].

As we can see from Fig. (**3**) [3], two regions of interest are formed: one near the gate metal (region I) and the second region II near the heterointerface; let us consider which current phenomena occur in these regions.

Region I: the electron collected on the metal surface (its Fermi level is initially lower than that of AlGaAs semiconductor) produce a negative electric field $\mathcal{E}_s$ [3]. Suppose the AlGaAs film is of n-type conductivity. Then the positively charged ionized donors in this layer produce a positive field gradient counteracting $\mathcal{E}_s$, and at some distance d_1 completely annuls it to produce equilibrium [3]. Thus, the energy bands in the structure components are bending and cause the alignment of the Fermi levels, as illustrated by Fig. (**3**) [3]. The potential energy of the electron in this region measured from the Fermi level is [3]:

$$|e|V_I(z)=|e|(V_s-V_g)+|e|\mathcal{E}_s z+(|e|^2/\epsilon_2)\int_0^z dz'\int_0^{z'} [N_d(z'')-n(z'')]dz'' \qquad (4.1)$$

We have to note: the direction normal to the interfaces has been chosen as the z-direction, the x-direction being chosen along the length of the device [3], $N_d(z)$ and n(z) are the doping and electron concentrations, respectively. The first term in the expression (4.1) is the potential at $z = 0$ appeared due to the surface barrier potential V_s and the applied gate voltage V_g [3]. The term $|e|\mathcal{E}_s z$ takes care of the potential drop at some distance z due to the surface field $\mathcal{E}_s$ [3]. The third term stands for the increase caused by the charges in the layer, *e* is the elemental charge (here the charge of the electron) and ϵ_2 is the permittivity of AlGaAs material [3]. Let us assume that the AlGaAs film is doped uniformly up to the distance d_d, but not doped for another length d_u adjacent to the heterointerface (see Fig. (3) [3]). We can simplify the expression (4.1) by neglecting *n(z)*:

$$V_I(z) = V_s - V_g + \mathcal{E}_s z + az^2 \tag{4.2}$$

where [3]:

$$a = |e| N_d / 2\epsilon_2 \tag{4.3}$$

Thus, the electric field in this region is as follows [3]:

$$\mathcal{E}_I = \mathcal{E}_s + 2az \tag{4.4}$$

Note that the distance d_1 is the distance of equilibrium, where $\mathcal{E}_I = 0$ and $V_I(d_1) = E_{F2}$ [3]. Under these conditions [3]:

$$\mathcal{E}_s = -2ad_1 \tag{4.5}$$

The region II (region of depletion near the heterointerface) is formed due to the electron migration from the AlGaAs layer to the GaAs layer. The conduction band edge of the GaAs material is lower in energy by ΔE_c. Suppose this region extends to the distance d2. Then the field and the potential in this region are to be estimated for two subregions [3]:

1)   for $d_2 < z < d_d$:

$$\mathcal{E}_{II}(z) = 2a(z - d_2) \tag{4.6}$$

$$|e| V_{II}(z) = E_{F2} + |e| a(z - d_2)^2 \tag{4.7}$$

2)   for $d_d < z < d$:

$$\mathcal{E}_{II}(z) = 2a(d_d - d_2) \tag{4.8}$$

$$|e|V_{II}(z) = E_{F2} + |e|a(d_d\text{-}d_2)[2z\text{-}(d_d\text{+}d_2)] \qquad (4.9)$$

To compute the potential distribution in GaAs film we have to take into account the following boundary conditions [3]:

i)   the potential energy decreases by the discontinuity in the conduction band edges, ΔE_c, at the interface;

ii) there are no surface states at the interface and the displacement is continuous across the interface;

iii) the GaAs layer is completely undoped.

Regarding these boundary conditions, we obtain the field in GaAs film as follows [3]:

$$\mathcal{E}(z) = (\epsilon_2/\epsilon_1)2a(d_d\text{-}d_2)\text{-}(|e|/\epsilon_1)\int_d^z n(z)dz \qquad (4.10)$$

where $n(z)$ and ϵ_1 are the electron concentration of GaAs and permittivity of the material, respectively.

As is seen, the field decreases with the distance; the corresponding potential energy increases from its value at the interface to the bulk value at some distance d_∞, where the field becomes zero [3]:

$$\int_d^\infty n(z)dz = N_d(d_d\text{-}d_2) \qquad (4.11)$$

or when the total accumulated charge is equal to the depletion charge [3].

The potential distribution may be altered by changing the gate voltage. However, the gate voltage changes only the width of the depletion region in the metal side so long as d_1 and the whole GaAs layer is depleted [3].

The threshold voltage for the charge control V_{th1} is obtained by eliminating $d_1 = d_2$ from Eqs. (4.2) and (4.7) and using the following relation (Fig. (**3**)) [3]:

$$|e|V_{II}(d) - \Delta E_c + E_{F1} = 0 \tag{4.12}$$

The threshold voltage is written as follows [3]:

$$V_{th1} = V_s - E_{F2}/|e| - a(d - d_0)^2 \tag{4.13}$$

where,

$$a(d_0)^2 = a(d_U)^2 + (\Delta E_c - E_{F2} - E_{F1})/|e| \tag{4.14}$$

The charge accumulated in the GaAs layer for operating conditions may be derived from the field produced at the interface under the complete depletion regime of the AlGaAs layer [3]. The field and the potential are expressed by the next equations [3]:

$$\mathcal{E}(d) = \mathcal{E}_s + 2ad_d \tag{4.15}$$

$$V(d) = V_s - V_g + \mathcal{E}_s d - a(d_d)^2 + 2add_d \tag{4.16}$$

The surface field $\mathcal{E}_s$ can be evaluated with help of Eqs. (16) and (12) [3]:

$$\mathcal{E}_s = (1/d)(V_g - V_s + \Delta E_c/|e| - E_{F1}/|e| + a(d_d)^2) - 2ad_d \tag{4.17}$$

As the next step we obtain [3]:

$$\mathcal{E}(d) = (1/d)(V_g - V_s + \Delta E_c/|e| - E_{F1}/|e| + a(d_d)^2) \tag{4.18}$$

The accumulated areal charge density is [3]:

$$Q_s = (\epsilon_2/d)(V_g - V_s + \Delta E_c/|e| - E_{F1}/|e| + a(d_d)^2) \tag{4.19}$$

The Fermi level E_{F1}, however, depends on the surface electron density n_s [3]:

$$E_{F1} = E_{F0} + \alpha n_s \tag{4.20}$$

where $\alpha = 1.25 \times 10^{-17}$ eV·m^2 for the GaAs/AlGaAs system [3]. Therefore, the accumulated areal charge density n_s is as follows [3]:

$$|Q_s| = (\epsilon_2/d_{eff})(V_g - V_{th2}) \tag{4.21}$$

Here [3]:

$$V_{th2} = V_s - \Delta E_c / |e| - E_{F0} / |e| - a(d_d)^2 \tag{4.22a}$$

$$d_{eff} = d + \epsilon_2 \alpha / e^2 \tag{4.22b}$$

Now we have to calculate the current-voltage characteristic of the HEMT.

4.2. Current-Voltage Characteristic [3]

We suppose that a voltage V_d is applied between the drain and the source, and a gate voltage V_g is applied between the gate with a length L and a width w, and the source. The zero of the x-coordinate is a point where the gate starts. The bias of interest is V(x) at a distance l_x from the gate end [3].

Then, the effective gate voltage as a function of the distance x is [3]:

$$V_{eff} = V_g - V(x) \tag{4.23}$$

and the surface charge density as a function of x is [3]:

$$|e|n(x) = (\epsilon_2 / d_{eff})(V_{eff} - V_{th2}) \tag{4.24}$$

The current flowing through the sample L can be expressed as [3]:

$$I = \mu(\mathcal{E})|e|n(x)[dV(x)/dx]w \tag{4.25}$$

where $\mu(\mathcal{E})$ is the electron mobility for a field [3]:

$$\mathcal{E} = -\, dV(x)/dx \tag{4.26}$$

To derive the expression for the current-voltage characteristic we will use a simple model describing the velocity-field characteristic [3].

Suppose that the mobility has a field-independent value μ_0 valid up to a threshold field $\mathcal{E}_{th}$, beyond which the velocity is assumed to have a constant value v_s [3]. Integrating (4.3) gives [3]:

$$V(x) = V_g - V_{th2} - \{[V_g - V_{th2} - V(0)]^2 - (2Id/\epsilon_2\mu_0 w)x\}^{1/2} \tag{4.27}$$

Here $V(0) = R_sI$, R_s being the resistance of the channel between the source contact and the front end of the gate [3]. The bias across the length of the gate is [3]:

$$V(L)=V-(R_d+R_s)I=V_g-V_{th2}-[(V_g-V_{th2}-R_sI)^2-(2/\epsilon_2\mu_0w)ILd]^{1/2} \tag{4.28}$$

where R_d is the resistance of the channel between the drain contact and the rear end of the gate [3].

Now we obtain saturation current I_s by evaluating the field $(-dV(x)/dx)$ from the equation (4.26) and putting it equal to $\mathcal{E}_{th}$ [3]:

$$I_s=(\epsilon_2wv_s/d)\{[V_g-V_{th2}-(V(0))^2+(\mathcal{E}_{th})^2L^2]^{1/2}-\mathcal{E}_{th}L\} \tag{4.29}$$

This expression for the saturation current I_s may be simplified for large values of the length L [3]:

$$I_s=C(v_s/L)(V_g-V_{th2}-V(0))^2/2\mathcal{E}_{th}L \tag{4.30}$$

$$C=\epsilon_2wL/d \tag{4.31}$$

and the saturation current is proportional to the square of the effective voltage [3].

For the small values of L the saturation current I_s can be approximated as follows [3]:

$$I_s=C(v_s/L)(V_g-V_{th2})(1+R_sCv_s/L)^{-1} \tag{4.32}$$

The intrinsic transconductance g_m is [3]:

$$g_m=C(v_s/L)(1+R_sCv_s/L)^{-1} \tag{4.33}$$

One has to note that for enhanced performance the gate length is required to be small, and the saturation velocity should be large [3]. The source resistance should also be small. The constant C is the capacitance between the gate metal and the quantum well layer. The current gain (as one of the main characteristics of the transistor) is therefore given by [3]:

$$I_s/|I|=g_m/wC=f_T/f \tag{4.34}$$

where

$$f_T = (1/2\pi)(g_m/C)$$ **(4.35)**

Thus, this simple model makes it possible to derive basic parameters limiting the performance of the HEMT [3].

5. METAL-OXIDE-SEMICONDUCTOR FIELD EFFECT TRANSISTOR [4]

Metal-oxide-semiconductor field effect transistors (MOSFETs) (Fig. (**4**)) are belonging to the group of the most important devices of the microelectronics.

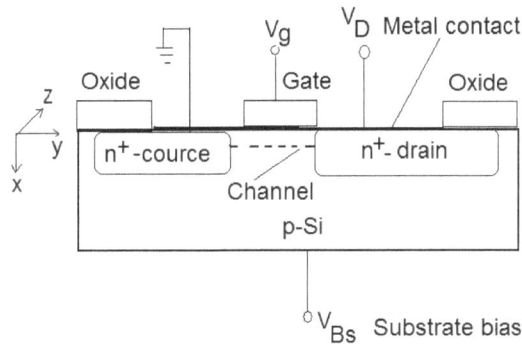

Figure 4: A generalized scheme of the common MOSFET [4].

The versions of the field effect transistors (FETs) differ according to the type of channel carriers. We have *n*-channel and *p*-channel devices, the *n*-channels are formed by electrons and are more conductive, while p-channels are formed by holes and are more conductive with more negative gate bias [4]. FETs are called enhancement-mode, or normally-off, if at zero gate bias the channel conductance is very low and one must apply a gate voltage to form a conductive channel [4]. The counterpart is called depletion-mode, or normally-on, when the channel is conductive with zero gate bias and we must apply a gate voltage to form the transistor off [4].

A channel can be formed by a surface inversion layer. The surface inversion channel is a two-dimensional (2D) charge sheet of the thickness in the order of 5 nm [4].

Now we consider a common MOSFET: it is a four-terminal device that consists of p-type semiconductor substrate into which two n^+-regions (the source and drain) are formed usually by ion implantation [4]. The SiO_2 gate dielectric is prepared by thermal oxidation of Si for a high-quality SiO_2-Si interface [4]. The metal contact on the insulator is called the gate; heavily doped polycrystalline silicon is more commonly used as the gate electrode [4]. The basic parameters of the device are the following: the length of the channel L (there is a distance between the two metallurgical n^+-p-junctions), the width of the channel z, the insulator thickness d, the depth of the junction r_j, and the substrate doping level N_A [4].

How does the device operate? When ground or a low voltage is applied to the gate, the main channel is shut off, and the source-to-drain electrodes correspond to two p-n-junctions connected back to back [4]. When a sufficiently large positive bias is applied to the gate so that a surface inversion layer (or channel) is formed between the two n+-regions, the source and the drain are then are connected by a conducting surface n-channel [4]. Through this channel a large current can flow. The conductance of this channel can be modulated by varying the gate voltage [4]. The back surface contact (or substrate contact) can be at the reference voltage or reverse biased; this substrate voltage will also affect the channel conductance [4].

The characteristics of the space charge under the non-equilibrium condition are derived under two assumptions [4]: (1) the majority-carrier quasi-Fermi level E_{Fp} is the same as that of the substrate and it does not vary with distance from the bulk to the surface (constant with x), and (2) the minority-carrier quasi-Fermi level E_{Fn} is lowered by the drain bias by an amount dependent on the y-position.

The most important models used for describing the function of the device (*i.e.*, the current-voltage characteristics) are the following: constant mobility, the charge sheet model, velocity-field relationship, field-dependent mobility, and the ballistic transport [4]. Now consider the sheet-charge model and the ballistic transport model.

The starting equation for a charge sheet model is an equation for the surface potential $\psi_s(y)$ at the drain under the onset of strong inversion [4]:

$$\psi_s(inv) \approx V_D + 2\psi_B \qquad (5.1)$$

Under strong-inversion conditions the inversion layer can be treated as a charge sheet with zero thickness ($x_i = 0$) [4]. Thus, the potential drop across this charge sheet is also zero. From Gauss' law, the boundary conditions on both sides of the charge sheet are [4]:

$$\mathcal{E}_{ox}\mathcal{C}_{ox} = \mathcal{E}_{s}\mathcal{C}_{s} - Q_n \tag{5.2}$$

To express $Q_n(y)$ throughout the channel, the surface potential is to be generalized from Eq.(5.1) to [4]:

$$\psi_s(y) \approx \Delta\psi_i(y) + 2\psi_B \tag{5.3}$$

where $\Delta\psi_I$ is the channel potential with respect to the source end:

$$\Delta\psi_i(y) \equiv [E_i(x{=}0, y{=}0) - E_i(x{=}0, y)]/q \tag{5.4}$$

and is equal to V_D at the drain end [4]. One has to note that the electric fields can be expressed as [4]:

$$\mathcal{E}_{ox} = (V_G - \psi_s)/d = (V_G - (\Delta\psi_i + 2\psi_B))/d \tag{5.5}$$

$$\mathcal{E}_s = [2qN_A((\Delta\psi_i + 2\psi_B)/\mathcal{C}_s]^{1/2} \tag{5.6}$$

In Eq. (5.5) an ideal MOS system with zero work function is assumed [4]. Eq. (5.6) is simply the maximum field at the edge of the depletion region [4]. Combining Eqs. (5.1) - (5.6) and using the expression $C_{ox} = \mathcal{C}_{ox}/d$ we obtain [4]:

$$|Q_n(y)| = [V_G{-}\Delta\psi_i(y){-}2\psi_B]C_{ox}{-}\{2q\mathcal{C}_sN_A[\Delta\psi_i{+}2\psi_B]\}^{1/2} \tag{5.7}$$

To obtain the current-voltage characteristics (IVC) of the common MOSFET we will use the following idealized conditions [4]: (1) the gate structure corresponds to an ideal MOS capacitor; it means that there are no interface traps for mobile oxide charge; (2) only drift current will be taken into account; (3) the channel is uniformly doped; (4) reverse leakage current is negligible; (5) the transverse field ($\mathcal{E}_x$ in the x-direction) in the channel is much larger than the longitudinal field ($\mathcal{E}_y$ in the y-direction). These conditions are called a gradual-channel approximation [4]. The condition (1) implies that the requirements of zero fixed oxide charge and work-function difference are removed, and their effects are included in a flat-band voltage V_{FB} required by the gate to produce the flat-band condition [4]. Consequently, V_G is replaced by V_G-V_{FB} for the inversion charge, giving [4]:

$$|Q_n(y)|=[V_G\text{-}V_{FB}\text{-}\Delta\Psi_i(y)\text{-}2\Psi_B]C_{ox}\text{-}\{2\varepsilon_s qN_A[\Delta\Psi_i(y)+2\Psi_B]\}^{1/2} \qquad (5.8)$$

Under such idealized condition, the channel current at any y-position is given by [4]:

$$I_D(y)=z|Q_n(y)|v(y) \qquad (5.9)$$

where $v(y)$ is the average carrier velocity [4]. Since the current has to be continuous and constant throughout the channel, integration of Eq.(5.9) from 0 to L gives [4]:

$$I_D=zL^{-1}\int_0^L |Q_n(y)|\,v(y)dy \qquad (5.10)$$

The carrier velocity $v(y)$ is the function of y-position since the longitudional field $\mathcal{E}_y(y)$ is variable [4]` For shorter channel lengths (as in novel devices),higher field causes, among other effects, the ballistic transport [4]. Below we describe its main features.

The ballistic transport of charge carriers is realized in ultra-short channel lengths where dimensions are on the order of or shorter than the mean free path; in this case channel carriers do not suffer from scattering [4]. They can gain energy from the applied electric field without losing it to the lattice through scattering, and can acquire a velocity much higher than the saturation velocity [4]. This effect is called ballistic transport.

The ballistic transport is important since it points out that the current and transconductance can be higher than that of saturation velocity, giving an additional incensitive for shrinking the channel length [4].

To analyze the saturation current in the ballistic regime we apply the Eq.(10) to the maximum-potential point. We are starting with the generalized form [4]:

$$I_{Dsat} = z|Q_n|v_{eff} \qquad \qquad (5.11)$$

where $|Q_n|$ has the maximum value at the source as $C_{ox}(V_G-V_T)$, and v_{eff} is an effective average carrier velocity, V_T is a threshold voltage [4].

The $v_{eff\ max} \equiv v_{th}[=(2k_BT/m^*)^{1/2}]$, according to classical thermal equilibrium, v_{th} is the thermal velocity [4]. The detailed examination of the system shows that for higher inversion charge density, the random velocity can exceed this thermal limit [4]. This is a quantum-mechanical effect, called carrier degeneracy, where the mean carrier energy is pushed to a higher state than the thermal energy. This higher value is called the injection velocity v_{th}, related to the Fermi energy with respect to the quantized energy E_n inside the potential well where carriers reside [4]:

$$v_{inj}=(2k_BT/\pi m^*)^{1/2}\{F_{1/2}[(E_F-E_n)/k_BT]\}/\ln\{1+\exp[(E_F-E_n)/k_BT]\} \qquad (5.12)$$

where $F_{1/2}$ is the Fermi-Dirac integral [4]:

$$F_{1/2}[(E_F-E_n)/k_BT] = \int_{E_n}^{\infty} \{[(E-E_n)/k_BT]^{1/2}/\{1+\exp[(E-E_n)/k_BT]\}(dE/k_BT) \qquad (5.13)$$

With a small inversion charge, Eq.(5.12) reduces to $(2k_BT/m^*)^{1/2}$, and $v_{inj}=v_{th}$ [4]. If the inversion charge is high, Eq.(5.12) can be simplified to [4]:

$$v_{inj} = (8h/3m^*)[|Q_n|/2\pi q]^{1/2} = (8h/3m^*)[C_{ox}(V_G-V_T)/2\pi q]^{1/2} \qquad (5.14)$$

This expression is a function of the inversion charge.

The saturation current of (5.11) can be rewritten as [4]:

$$I_{Dsat} = r_nz|Q_n|v_{inj} = (8r_nh/6\pi m^*)[C_{ox}(V_G-V_T)^{3/2}]/(2\pi q)^{1/2} \qquad (5.15)$$

where $r_n = v_{eff}/v_{inj}$ is the index of ballisticity [4]. In the extreme of ballistic transport, $r_n = 1$, and it sets the ultimate current drive for $L\rightarrow0$ [4]. The transconductance is given by [4]:

$$g_m = (4r_nzh/2\pi m^*)[C_{ox}(V_G-V_T)/2\pi q]^{1/2} \qquad (5.16)$$

Here it is seen that both I_{Dsat} and g_m are independent of channel length L.

The index of ballisticity is also interpreted by back scattering R of channel carriers at the drain back to the source [4]. Furthermore, since mobility is also a consequence of scattering, there should be some relationship between r_n and the low-field mobility μ_n [4]:

$$r_n = (v_{eff}/v_{inj}) = (1-R)/(1+R) = [(v_{inj})^{-1} + (\mu_n(\mathcal{E}(0^+))^{-1}]^{-1} \qquad (5.17)$$

where $\mathcal{E}(0^+)$ is the field of a potential $k_B T$ down from the maximum toward the drain [4].

The next part of this lecture presents a very interesting invention based on fundamentals described above.

6. SPIN INJECTION DEVICE HAVING SEMICONDUCTOR-FERROMAGNETIC SEMICONDUCTOR STRUCTURE AND SPIN TRANSISTOR [5]

The semiconductor industry is increasingly being driven to decrease the size of semiconductor devices located on integrated circuits (ICs). For example, miniaturization is needed to accommodate the increasing density of ICs necessary for today's semiconductor products. Increased packing density and device size reduction has forced semiconductor device structures such as transistors to be located ever closer to one another [5].

As semiconductor device components become located closer together, the problem of so-called Joule heating becomes more pressing. In general, bulk flow of electrons within conventional semiconductor devices generates heat that must be dissipated. The problem of Joule heating is limiting the ability of semiconductor manufacturers to satisfy the demand for even smaller, more compact devices. Manufacturing smaller devices using known charge diffusion technologies results in increased Joule heating [5].

One potential solution to the Joule heating problem is to utilize the spin states of electrons rather than the charge of electrons. In addition to having a charge, electrons also have a well defined spin. "Spin" is a property of an electron that is generally related to the angular momentum of an electron about an axis within the electron. An electron has to spin states—spin up ($+\frac{1}{2}$) and spin down ($-\frac{1}{2}$). These discernable spin states can be flipped or toggled for purposes of identifying logic "0" and "1" values [5].

The amount of energy required to alter the electron spin may be less than the amount of energy needed for bulk charge movement (as is done in traditional semiconductor devices). For this reason, spin-based devices may offer a promising modality for very small semiconductor-based devices and provide the potential for faster logic devices, such as field-effect transistors (FETs), and may consume less power and generate less heat [5].

One significant challenge to the realization of spin-based FETs is the ability to electronically inject spin-polarized charge carriers (*e.g.*, electrons) into a suitable substrate (*e.g.*, single crystalline silicon) or semiconductor channel at room or ambient temperature. Spin-polarized refers to the state in which all or substantially all of the electrons are initialized to a given spin state (*e.g.*, spin "up" or spin "down" state) [5].

One known manner of initializing or polarizing electrons to have a certain spin state is based on passing the electrons or holes through ferromagnetic materials (which are metals) that have been magnetized then into semiconductor materials. More particularly, in certain known spin devices, magnetic forces spin polarize electrons as they pass through a ferromagnetic material. The spin-polarized electrons pass from a ferromagnetic material into a semiconductor-based material. Unfortunately, efficient spin injection using these types of structures may not be achievable due to the conductivity mismatch between the ferromagnetic material and the semiconductor-based material. More particularly, these effects cause electrons that were spin-polarized in the ferromagnetic material to randomize and assume various spin states when they are injected into the semiconductor material. This randomization negates or reduces the spin polarization that was achieved using ferromagnetic material, thereby making it difficult or impossible to achieve a common and detectable spin state [5].

Another known manner of initializing spin of electrons is to inject spin from a dilute magnetic semiconductor that serves to align spin in the presence of a magnetic field. Such devices may operate well at low temperatures but are not suitable at room temperature due to the magnetic semiconductor materials losing their spin-aligning capabilities at room temperatures.

Yet another known manner of initializing electrons to have a certain spin is based on quantum mechanical tunneling and use of an intermediate layer of silicon dioxide. Tunnel injection is, however, associated with high resistance, which is detrimental to FET operations [5].

A further known spin initialization method relies on optical polarization of electrons. This technique, however, has proved difficult, and it is generally believed to be incompatible or difficult to effectively implement with most microelectronic devices [5].

Authors [5] refer readers to various classical works describing a spin injection device and a FET having a ferromagnetic-semiconductor-ferromagnetic structure. A spin-polarized carrier is injected into the channel region from the ferromagnetic source and detected from the ferromagnetic drain. A device having such a structural configuration, however, may suffer from spin randomization [5].

Thus there is a need for spin injection devices, spin FETs that are capable of efficiently injecting spin-polarized electrons into a substrate, such as a single crystalline silicon material or substrate. Such devices, FETs and methods should be able to spin-polarize a substantial number of electrons to a particular spin state without spin alignment randomization associated with interfacial effects between ferromagnetic and semiconductor materials. Such spin injection devices should also have low resistance, which is an important figure of merit for the overall FET performance, particularly in terms of the power consumed by the device. In addition, such spin injection devices, FETs and methods should be capable of fabrication using accepted fabrication systems techniques and be amenable to incorporation into current and contemplated microelectronic devices. Further, such devices and FETs should be operable at room or ambient temperatures so that they can be used in various commercial devices and applications without environmental limitations. It would also be desirable to have spin FETs that can serve as an alternative to known silicon CMOS devices that are based on charge diffusion [5].

A spin injection device proposed in [5] includes a first semiconductor material, a second semiconductor material that is different from the first semiconductor material, and a spin-polarizing ferromagnetic material interposed between the first and second semiconductor materials. Charge carriers from the first semiconductor material are spin-polarized upon traversing the ferromagnetic material, and then injected into the second semiconductor material [5].

Another model [5] is a spin injection device including a first semiconductor material, a second semiconductor material, and a spin-polarizing ferromagnetic material interposed between the first and second semiconductor materials. The first semiconductor material has an amorphous or polycrystalline structure, and the second semiconductor material has a single crystalline structure. The device is configured so that a first Schottky barrier formed by the first semiconductor material and the spin-polarizing ferromagnetic material is higher than a second Schottky barrier height formed by the spin-polarizing ferromagnetic material and the second semiconductor material, and charge carriers having random spin from the first semiconductor material traversing the spin-polarizing ferromagnetic material are spin-polarized, and spin-polarized charge carriers being injected from the spin-polarizing material into the second semiconductor material [5].

In an alternative embodiment of the invention [5], a spin transistor includes a source, a drain, a channel electrically connecting the source and drain, and a gate configured for controlling conduction of spin-polarized elements through the channel. The source is a spin injection device that includes a first semiconductor material, a second semiconductor material that different than the first semiconductor material, and a spin-polarizing ferromagnetic material interposed between the first and second semiconductor materials. Charge carriers from the first semiconductor material are spin-polarized upon traversing the ferromagnetic material, and then injected into the second semiconductor material [5].

A spin field-effect transistor proposed by authors [5] includes a source, a drain, a channel electrically connecting the source and the drain and a gate configured for controlling conductivity of spin-polarized elements through the channel. The source is a spin injection device that includes a first semiconductor material, a second semiconductor material, and a spin-polarizing ferromagnetic material interposed between the first and second semiconductor materials. The first semiconductor material has an amorphous or polycrystalline structure, and the second semiconductor material has a single crystalline structure. The device is configured so that a height of a first Schottky barrier formed by the first semiconductor material and the spin-polarizing ferromagnetic material is larger than a height of a second Schottky barrier height formed by the spin-polarizing ferromagnetic material and the second semiconductor material, and charge carriers having random spin from the first semiconductor material traversing the spin-polarizing ferromagnetic material are spin-polarized, and spin-polarized charge carriers being injected from the spin-polarizing material into the second semiconductor material [5].

In some structures described in [5], the first and second semiconductor materials have different crystalline structures. The first semiconductor material, for example, may be amorphous or polycrystalline (*e.g.*, amorphous or polycrystalline silicon), and the second semiconductor material, for example, may be single crystalline (*e.g.*, single crystalline silicon).

Further, rather than a single element material, the semiconductor materials can be compounds including one or more elements. For example, the first semiconductor material may be polycrystalline or amorphous AlGaAs, GaAs or GaN, and the second semiconductor material may be single crystalline GaAs, AlGaAS, GaN, InSb or InP, provided that the Schottky barrier height between the first semiconductor and the ferromagnetic material is significantly higher than that between the ferromagnetic material and the second semiconductor [5].

The first semiconductor material may instead be a single crystalline material, and the second material may be an amorphous or polycrystalline material. For example, the first semiconductor material may be single crystalline GaAs, AlGaAS, GaN, InSb or InP, and the second semiconductor material, is polycrystalline or amorphous AlGaAs, GaAs or GaN, provided the Schottky barrier height between the first semiconductor and the ferromagnetic material is significantly higher than that between the ferromagnetic material and the second semiconductor [5].

Additionally, the first and second semiconductor materials may be different materials having the same crystalline structure provided the Schottky barrier height between the first semiconductor and the ferromagnetic material is significantly higher than that between the ferromagnetic material and the second semiconductor. For example, the first semiconductor material is single crystalline AlGaAs, and the second semiconductor material is single crystalline GaAs [5].

A thickness of the first semiconductor material may be greater than a thickness of the first semiconductor material required for a non-zero quasi-neutral region, and the ferromagnetic material may be a permalloy film (or other ferromagnetic materials such as Fe, Co, Ni, *etc.*) having a thickness of about 1-100 nm, and a spin diffusion length (L_{SD}) that is significantly shorter than an electronic energy relaxation mean free path (L_E). For example, the ferromagnetic material may have a thickness of at least two times a spin diffusion length (L_{SD}) of the spin-polarizing ferromagnetic material [5].

A spin injection device [5] may also be a source of a field effect transistor, such as a spin metal-oxide-semiconductor field-effect transistor. A material of a channel of a field effect transistor may be the same material as the second semiconductor material into which spin-polarized charge carriers are injected. Given the structure of embodiments, they can operate at ambient temperature, and spin-polarization can be achieved without having to modulate magnetization of a ferromagnetic material since the ferromagnetic material is subject to a substantially constant magnetization level, *e.g.*, saturation [5].

The structures proposed by researchers [5] provide a new approach for efficient spin injection and spin FETs that utilize a hybrid semiconductor-ferromagnetic-semiconductor spin injection structure as opposed to known ferromagnetic-semiconductor-ferromagnetic structures and BJT spin valve transistors. Embodiments may operate at room temperature and may be made using accepted photolithography methods rather than direct bonding. Embodiments are advantageously structured to inject spin-polarized charge carriers into a substrate or semiconductor material, for example, single crystalline silicon, thus allowing introduction and manipulation of spin-polarized charge carriers by other microelectronic devices, while eliminating or mitigating Joule heating issues associated with known devices based on charge diffusion [5].

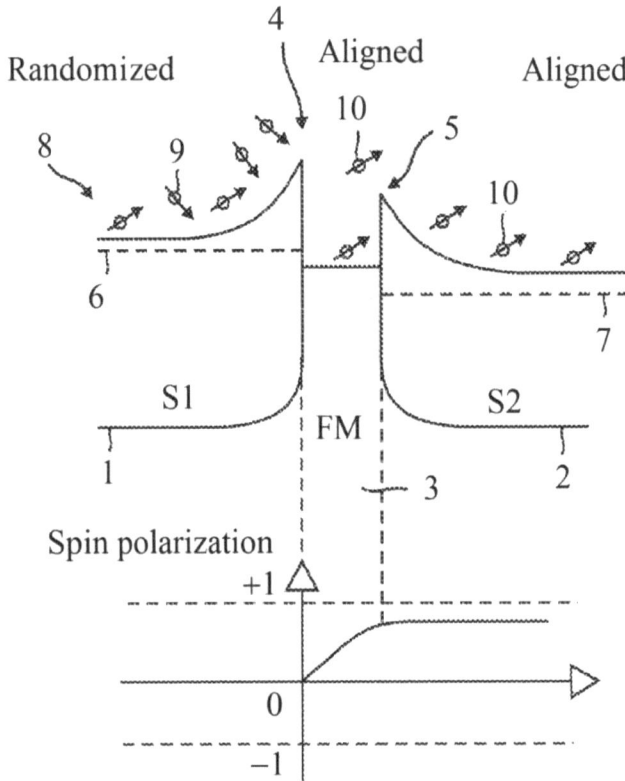

Figure 5: An energy band/spin polarization diagram of a spin injection device: 1 – first semiconductor material, 2 – second semiconductor material, 3 – ferromagnetic material, 4 – first Schottky barrier, 5 - second Schottky barrier, 6 – first Fermi level, 7 – second Fermi level, 8 – charge carriers, 9 – random spin, 10 – spin-polarized carriers [5].

Referring to Figs. (**5-8**) [5], a spin injection or initialization device constructed in accordance with one embodiment includes a first semiconductor material (**S1**), a second semiconductor material (**S2**) that is different than the first semiconductor material, and a spin-polarizing ferromagnetic material (**FM**). The ferromagnetic material **FM** is interposed between the first and second semiconductor materials [5].

As shown in Fig. (**5**) [5], embodiments utilize a structural configuration including a first Schottky barrier or junction (generally referred to as "first Schottky barrier) defined or formed by the contact between first semiconductor material and the ferromagnetic material and a second Schottky barrier or junction (generally referred to as "second Schottky barrier") defined or formed by the contact between the ferromagnetic material and the second semiconductor material. The first Schottky barrier has an associated first Fermi energy level (represented as dotted line), which is higher than a second Fermi energy level (represented as dotted line) associated with the second Schottky barrier. In these structures the first Schottky barrier is higher than the second Schottky barrier when, for example, the ferromagnetic material is under sufficient magnetization [5]. According to one embodiment, the ferromagnetic material is subject to a substantially constant level of magnetization, *e.g.*, saturation, or the maximum induced magnetic moment or magnetization that can be obtained in a given magnetic field [5].

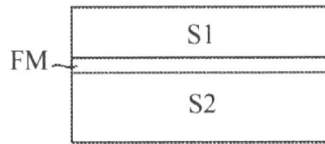

Figure 6: A cross-sectional view generally illustrating components of a spin injection device [5].

With this structural configuration, charge carriers (*e.g.*, electrons or holes) (generally referred to as electrons for ease of explanation) from the first semiconductor material may have random spin, and are spin-polarized as they traverse the ferromagnetic material. Spin-polarized electrons are then injected from the ferromagnetic material into the second semiconductor material assuming they have sufficient kinetic energy. The second semiconductor material may serve as a substrate and source of spin-polarized electrons for other micro-electronic components [5].

As indicated by the "S1" and "S2" identifiers, the first and second semiconductor materials are different materials. The first and second semiconductor materials are different and define a structure in which the first Schottky barrier is higher than the second Schottky barrier. The material may also be different in that they have different energy band gaps, *i.e.*, the first semiconductor material (*e.g.*, amorphous or polycrystalline silicon) has an energy band gap that is wider than an energy band gap of the second semiconductor material (*e.g.*, single crystalline silicon) [5].

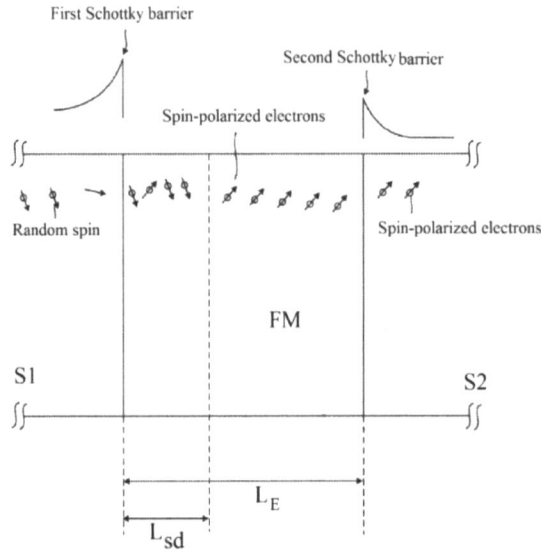

Figure 7: Spin polarization of charge carriers as they traverse an intermediate ferromagnetic material of a spin injection device [5].

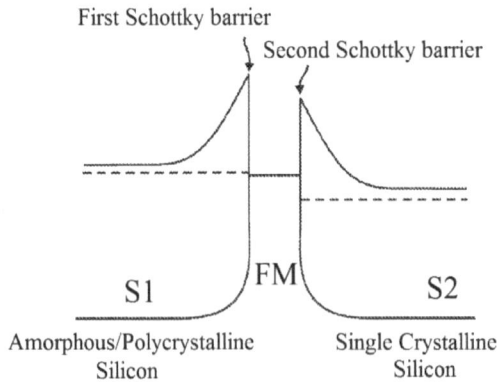

Figure 8: An energy band diagram of a spin injection device that includes a ferromagnetic material interposed between different crystalline forms of silicon [5].

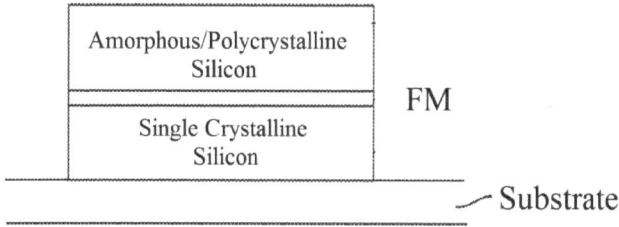

Figure 9: Cross-sectional view of a spin injection device constructed as shown in Fig. (**4**) [5].

According to other structures proposed by the authors [5], the semiconductor materials are different in that they have different compositions, compounds or elements, different chemical properties and/or crystalline properties. The first and second semiconductor materials may, for example, have different crystalline structures. In one embodiment, the first semiconductor material is amorphous or has a polycrystalline structure, and the second semiconductor material has a single crystalline structure. With this configuration, electrons that may initially have a random spin or that are not polarized are spin-polarized by the ferromagnetic material, and spin-polarized electrons are advantageously injected into the second or single crystalline semiconductor material. Substantially all of the spin-polarized charge carriers injected into the second semiconductor material S2 may maintain the common spin alignment [5].

Referring to Figs. (**8**) and (**9**) [5], a spin injection device constructed according to one embodiment includes first and second semiconductor materials that are different crystalline forms of silicon. The first semiconductor material is amorphous or polycrystalline silicon, and the second semiconductor material is single crystalline silicon, *e.g.*, formed from a single crystalline n^+-silicon wafer or substrate. With this configuration, electrons that may initially have a random spin or that are not polarized, traverse through the ferromagnetic material, and spin polarized electrons are injected into the single crystalline silicon [5].

A spin injection device that includes silicon having different crystalline structures is beneficial because amorphous silicon has a wider apparent energy band gap than single crystalline silicon, and single crystalline silicon material can be used for interfacing with microelectronic devices and substrates, which may also be silicon. This integration can be achieved without the need for direct bonding [5].

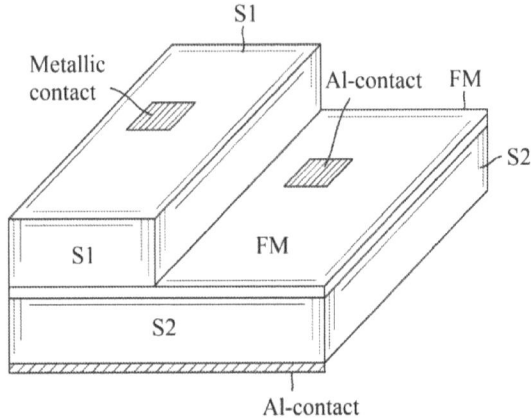

Figure 10: A perspective cross-sectional view of a spin injection device constructed as shown in Figs. (**8**) and (**9**) and having electrical contacts [5].

In another active element, the first semiconductor material is thicker than a thickness of the first semiconductor material that would be required for a non-zero quasi-neutral region. For example, the first semiconductor material in the form of amorphous or polycrystalline silicon can have a thickness of about 1-10.000 nm, and the second semiconductor material in the form of single crystalline silicon can have a thickness of about 10 nm to several mm. With this structure, the thickness of the ferromagnetic material can be about 1-100 nm [5].

Although Figs. (**8**) and (**9**) [5] illustrate an embodiment having amorphous silicon and single crystalline silicon; alternative embodiments can be implemented with other different semiconductor materials. For example, in other structures, the first semiconductor material **SM** may be one of polycrystalline or amorphous AlGaAs, GaAs or GaN, and the second semiconductor material may be one of single crystalline GaAs, AlGaAS, GaN, InSb or InP. The particular combination of compounds that is selected should result in a spin injection device structure in which the Schottky barrier height between the first semiconductor material and the ferromagnetic material is significantly higher than that between the ferromagnetic material **FM** and the second semiconductor material [5].

In other embodiments, the first semiconductor material may instead be a single crystalline material, and the second material may be an amorphous or polycrystalline material. For example, the first semiconductor material may be one of single crystalline GaAs, AlGaAS, GaN, InSb or InP, and the second semiconductor material may be one of polycrystalline or amorphous AlGaAs, GaAs or GaN, provided the Schottky barrier height between the first semiconductor and the ferromagnetic material is significantly higher than the Schottky barrier height between the ferromagnetic material and the second semiconductor [5].

In other structures [5], the first and second semiconductor materials are different materials or compounds having the same crystalline structure, provided that the Schottky barrier height between the first semiconductor and the ferromagnetic material is significantly higher than the Schottky barrier height between the ferromagnetic material and the second semiconductor. For example, the first semiconductor material may be single crystalline AlGaAs, and the second semiconductor material may be single crystalline GaAs. In this example, single crystalline AlGaAs has an energy band gap that is wider than that of single crystalline GaAs [5].

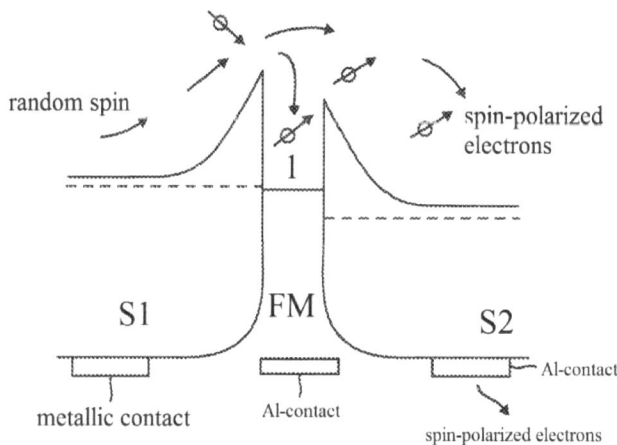

Figure 11: An energy band diagram of a spin injection device illustrating spin-polarized charge carriers injected into a second semiconductor material and other spin-polarized charge carriers that remain in the ferromagnetic material depending on electron energy levels; 1- spin-polarized electrons [5].

It should be understood, however, that the types of materials and combinations thereof discussed above are provided as examples of how embodiments can be implemented. Embodiments can be implemented using various combinations of amorphous/polycrystalline semiconductor materials, single crystalline materials, and various compounds). Exemplary criteria for the selection of suitable first and second semiconductor materials may include the Schottky barrier height being higher than the Schottky barrier height, the semiconducting materials S1, S2 having comparable electron concentrations, and the *k*-value of the conduction band minima between the semiconductor materials S1, S2 matching or substantially matching. The reciprocal space location of the electrons in the first semiconductor material **S1** should be nearly identical to that in the second semiconductor material so that the transport of electrons from the first semiconductor material to the second semiconductor material does not require the involvement of phonons [5].

The ferromagnetic material **FM** is preferably in the form of a thin film and can be, or include, for example, iron, cobalt, nickel and combinations thereof. The ferromagnetic material may include cobalt, but the ferromagnetic material FM should have a sufficiently short spin diffusion length. According to one embodiment, the ferromagnetic material is a permalloy film composed of about 80% nickel and about 20% iron and has an estimated spin diffusion length of about 6 nm), whereas the spin diffusion length of iron is about 2 nm, and the spin diffusion length of Cobalt is about 44 nm. Considerations for selecting the thickness of the ferromagnetic material so that spin-polarized electrons maintain sufficient kinetic energy are described in further detail below [5].

The ferromagnetic material preferably has a sufficiently short spin diffusion length (L_{SD}) (distance an electron diffuses through the ferromagnetic material before being spin polarized) and a thickness that allows electrons that are initially in a random spin state to traverse through the ferromagnetic material and be spin-polarized while maintaining most or substantially all of their kinetic energy. In this manner, the spin-polarized electrons have sufficient kinetic energy to overcome the second Schottky barrier and be injected into the second semiconductor material [5].

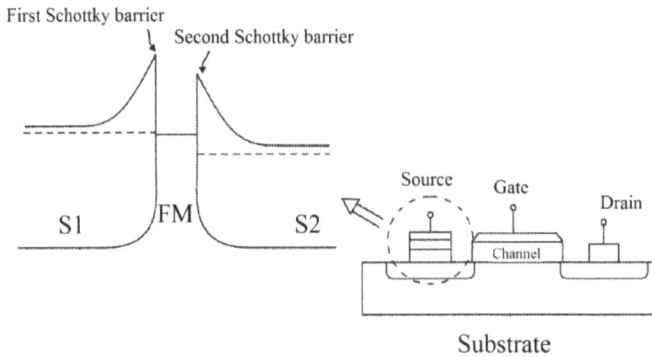

First Schottky barrier

Second Schottky barrier

S1   FM   S2

Source   Gate

Drain

Channel

Substrate

Figure 12: A field-effect transistor including a source in the form of a spin injection device structured as shown in Fig. (**5**) [5].

The thickness of the ferromagnetic material or film may depend on, for example, the type of ferromagnetic material, the spin diffusion length of the ferromagnetic material, the kinetic energy of the spin-polarized electrons upon entering the ferromagnetic material (which may depend on first Schottky barrier and the type of first semiconductor material), and the height of the second Schottky barrier (which may depend on the type of the second semiconductor material). In one embodiment, the ferromagnetic material is a permalloy having a thickness on the order of nanometers, *e.g.*, about 1-100 nm. Considerations for selecting the thickness of the ferromagnetic material so that spin-polarized electrons maintain sufficient kinetic energy are described in further detail below [5].

Referring again to Figs. (**5-7**) [5], electrons in the first semiconductor material **S1** have random spin, which can be spin up or spin down. Thus, the resulting spin polarization has a value of "0" on a scale from 0 to 1 as shown in polarization graph of Fig. (**5**) [5]. Electrons having a random spin are provided from the first semiconductor material **S1** (*e.g.*, amorphous or polycrystalline silicon) and traverse through the ferromagnetic material **FM** (*e.g.*, a permalloy), over the forward biased first Schottky barrier. With the ferromagnetic material being, *e.g.*, under saturation magnetization, random spin electrons are spin polarized as they are aligned with the magnetization of the ferromagnetic material. Electrons that were initially spin down are re-aligned to be spin up, and electrons that were initially spin up remain spin up (in the example in which the ferromagnetic material causes spin up polarization). Figs. (**5**) and (**7**) [5] illustrate an example in which electrons are aligned to a spin up state as they traverse through ferromagnetic material, however, it should be understood that embodiments can also be implemented using a spin down alignment with a different magnetization [5].

Electrons are spin polarized while traversing through the ferromagnetic material FM, and their kinetic energy remains essentially the same assuming that the electronic energy relaxation mean free path (L_E) (average distance an electron loses about 33% of its kinetic energy) is longer than L_{SD} (spin diffusion length). In this manner, a significant portion of electrons become spin-polarized while traversing the ferromagnetic material, and substantially all of the electrons, *e.g.* almost 100% of the electrons, can be spin-polarized if the thickness of the ferromagnetic material is more than about two times the spin diffusion length [5].

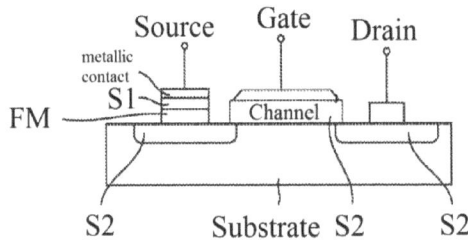

Figure 13: The field-effect transistor shown in Fig. (**12**) in further detail [5].

Since the kinetic energy of the spin-polarized electrons remains substantially the same, most of the spin-polarized electrons have sufficient kinetic energy to scale the second Schottky barrier and, therefore, are injected into the second semiconductor material (*e.g.* single crystalline silicon). Specifically, the kinetic energy of a spin-polarized electron upon exiting the ferromagnetic material must be greater than the height of the second Schottky barrier. In other words, the loss of electron kinetic energy due to an electron traversing the ferromagnetic material must less than the difference between the height of the first Schottky barrier and the height of the second Schottky barrier in order for the electron to be injected into the second semiconductor material [5].

Spin-polarized electrons (and any electrons that may not have been spin-polarized) that do not have sufficient kinetic energy to scale the second Schottky barrier are not injected into the second semiconductor material. Instead, these electrons remain within the ferromagnetic material. It is estimated that embodiments can be implemented so that about 50% to about 99% of the spin-polarized electrons will have sufficient kinetic energy and are injected from the ferromagnetic material **FM** into the second semiconductor material [5].

Fig. (**10**) [5] generally illustrates a perspective cross-sectional view of a spin injection device constructed in accordance with one embodiment. The spin injection device includes a ferromagnetic material interposed between different semiconductor materials S1 and S2, a metallic contact (*e.g.* aluminum) on the first semiconductor material to provide a source of charge carriers (electrons or holes), a contact (*e.g.* aluminum) on the second semiconductor material, and a contact (*e.g.*, aluminum) on the ferromagnetic material [5].

With further reference to Fig. (**11**) [5], spin-polarized electrons **1** within the ferromagnetic material FM that have lost sufficient kinetic energy and cannot overcome the second Schottky barrier remain in the ferromagnetic material and can be siphoned or drawn out of the ferromagnetic material through Al-contact (shown with reference to energy band structure illustrated in Fig. (**5**) [5]). Spin-polarized electrons that have sufficient kinetic energy and are injected into the second semiconductor material can be conducted through another Al-contact [5].

Referring to Figs. (**12**) and (**13**) [5], according to another embodiment, a spin injection or initialization device can be a component of a FET, such as a spin metal-oxide-semiconductor field-effect transistor (spin MOSFET). As generally illustrated in Fig. (**12**) [5], a FET typically includes a source, a gate, a drain, a channel or active region, and an underlying substrate or body. The channel is disposed under the gate and electrically connects the source and the drain. The gate controls conduction through the channel. The basic components and operation of FETs (based solely on charge diffusion) are well known and, therefore, are not described in further detail [5].

The FET proposed in [5] includes the spin injection or initialization device as the source. As shown in Figs. (**12**) and (**13**) [5], the channel may include or be composed of the same second semiconductor material. Thus, Figs. (**12**) and (**13**) [5] illustrate that embodiments can be integrated within FET devices to provide spin FETs. These devices may serve as a substitute for known FET devices that operate based on charge diffusion and may mitigate or eliminate issues of Joule heating associated with charge diffusion devices [5].

The active elements proposed by authors [5] can be fabricated using known and widely used photolithography fabrication equipment and processes and on a commercial scale. For example, beginning with a single crystalline silicon wafer or substrate (the second semiconductor material), the ferromagnetic material FM can be deposited on the single crystalline wafer, and then amorphous or polycrystalline silicon (the first semiconductor material S1) may be deposited on the ferromagnetic material. A metal layer may then be deposited on the amorphous or polycrystalline silicon to provide an electrical contact, *e.g.* a metallic contact (as shown in Fig. (**10**) [5]).

In this manner, embodiments advantageously provide efficient spin injection devices that can be fabricated using known deposition systems and techniques. Further, since the substrate or second semiconductor material may be single crystalline silicon, embodiments of spin-injection devices may be integrated with other microelectronic devices for use in various applications. Additionally, by use of such known fabrication methods, embodiments can be manufactured while maintaining purity of the ferromagnetic material or electron transport region, thus reducing or minimizing electron scattering and maintaining acceptable L_{SD} and L_E values [5].

For example, one method of fabricating active elements proposed in [5] having a first semiconductor S1-ferromagnetic **FM**-second semiconductor **S2** structure utilizes known photolithography techniques, programs and systems. Beginning with a $n+$-single crystalline silicon wafer or substrate (the second semiconductor material **S2**), a surface of the wafer may be cleaned and prepared, *e.g.*, using acetone, alcohol, dilute hydrogen fluoride (HF), de-ionized water, and N_2 blow drying. The single crystalline wafer may then be loaded into a deposition chamber. A layer of ferromagnetic material **FM** (*e.g.*, nickel having a thickness of about 5 nm), may then deposited on the single crystalline silicon wafer, followed by deposition of a layer of amorphous n-type silicon having a thickness of about 100 nm. A metallic layer (*e.g.*, aluminum) may then be deposited on the amorphous n-type silicon for use as an electrical contact (as generally illustrated in Fig. (**10**) [5]). These depositions may be performed, *e.g.*, sequentially in an e-beam evaporator. A layer of aluminum may also be deposited on the unpolished side of the single crystalline silicon wafer for use as an electrical Al-contact (*e.g.*, as generally illustrated in Fig. (**10**) [5]).

Photolithography may then be performed to define the area of a spin injection device, *e.g.*, using Mask **2** in L-edit design and suitable photolithographic parameters (*e.g.*, type of photoresist, spin speed, spin time, exposure time, developer mixtures and developing time. The exposed aluminum layer on the polished side may be etched, and KOH etching may be used to remove exposed amorphous silicon and to expose a layer of the nickel ferromagnetic layer. Lithography may then be performed to define an aluminum-nickel metallic contact utilizing Mask **3** in L-edit (or other photo mask layout software tools) design. Aluminum having a thickness of about 100 nm may be deposited for use as a Al-contact (as generally illustrated in Fig. (**10**) [5]) using the e-beam evaporator. The lift-off process may be completed using acetone and an ultra-sonic machine, and lithography may be performed to define mesa etch using Mask **1** in L-edit design. Aluminum etch may then used to remove exposed aluminum, and RIE etching may be performed for the mesa etch to remove single crystalline silicon. For example, embodiments can involve removing about 100 nm of amorphous silicon, about 5 nm of nickel ferromagnetic material, and some amount of the single crystalline silicon substrate. It should be understood, however, that spin injection device and spin FET embodiments can be fabricated using other methods and various photolithographic parameters [5].

Spin injection device and spin FET embodiments may have application in various devices including memory (*e.g.*, nonvolatile memory, "magnetic RAM"), reconfigurable logic architectures, sensors or detectors (*e.g.*, magnetic field sensors, position sensors, speed sensors, hard disk heads), and field programmable gate arrays. Embodiments of the invention, therefore, should not be limited, except to the following claims, and their equivalents [5].

ACKNOWLEDGEMENTS

Declared none.

CONFLICT OF INTEREST

The author(s) confirm that this chapter content has no conflict of interest.

REFERENCES

[1]   Snowden, C.M. *Introduction to Semiconductor Device Modeling*; World Scientific Publishing Co Pte Ltd.: Singapore, **1986**.

[2]   Ferry, D.K.; Goodnick, S.M. *Transport in Nanostructures*; Cambridge University Press: Cambridge, **2001**.

[3]   Nag, B.R. *Physics of Quantum Well Devices*; Kluwer Academic Publishers: Dordrecht, **2001**.

[4]   Sze, S.M.; Ng, Kwok K. *Physics of Semiconductor Devices*; Wiley Interscience: Hoboken, **2007**.

[5]   Xie, Ya-hong. Spin injection device having semiconductor-ferromagnetic semiconductor structure and spin transistor. US20100102319. **2010**.

Send Orders of Reprints at reprints@benthamscience.net

CHAPTER 4

Novel Spintronics Devices

Halyna Khlyap [*]

State Pedagogical University, 24 Franko str., UA-82100 Drohobych, Ukraine

Abstract: This lecture describes important spintronics devices: surface-spintronics device, spintronic transistor, spintronic devices with constrained spintronic dopants and other structures.

Keywords: Spintronics, transistor, spin polarization, spin states, surface spintronic device, spintronic dopants, magnetization, giant magnetoresistance.

1. INTRODUCTION

Spintronics is a novel branch of the contemporary microelectronics. Recently, all known devices and active elements have been used a charge as a carrier of the functional information. However, every charge carrier (the electron) has a spin. As we will see, the spin gives an additional degree of freedom to novel devices.

As is noted in [1], the core element of electronics is the transistor with its amplification, which relies on extremely pure semiconductor materials in which electrical conduction can be controlled by manipulating the carrier density using electric fields supplied by a gate. Semiconductors allow to precise tuning of carrier concentrations, band gap engineering, and, increasingly, also exhibit extremely long electron spin lifetimes [1]. Spin-electronics (or spintronics) based on semiconductors and (ferromagnetic)metal-insulator structures is therefore an active and promising field for both fundamental and practical applications [1].

The term spin stands for either the spin of a single electron $\mathbf{s}$, which can be detected by its magnetic moment $-g\mu_B\mathbf{s}$ (μ_B is the Bohr magneton and g is the electron g factor, in a solid generally different from the free-electron value of

*Address correspondence to Halyna Khlyap: State Pedagogical University, 24 Franko str., UA-82100 Drohobych, Ukraine; Tel: +49 631 414 4865; Email: gkhlyap17@yandex.ru

g_0=2.0023), or the average spin of an ensemble of electrons, manifested by magnetization [2]. The control of spin is then a control of either the population and the phase of the spin of an ensemble of particles, or a coherent spin manipulation of a single or a few spin systems [2].

Generation of spin polarization usually means creating a nonequilibrium spin population [2]. There are several ways to achieve the goal. As is known [2], traditionally spin has been oriented using optical techniques provided that circularly polarized photons transfer their angular momenta to electrons. However, for device applications electrical spin injection is more desirable [2]. In a electrical spin injection [2] a magnetic electrode (ferromagnetic metal) is connected to the sample (semiconductor, bulk or thin film, or insulator). When the current drives spin-polarized electrons from the electrode to the sample, nonequilibrium spin accumulates in the sample. The rate of spin accumulation depends on spin relaxation, the process of bringing the accumulated spin population back to equilibrium [2]. There are different mechanisms of spin relaxation: the most of them involve spin-orbit coupling to provide the spin-dependent potential, in combination with momentum scattering to provide a randomizing force. Typical time scales for spin relaxation [2] in electronic systems are measured in nanoseconds, while the range is from the picoseconds to microseconds. Spin detection typically relies [2] on sensing the changes in the signals caused by the presence of nonequilibrium spin in the system. The aim in the most electronic devices is to maximize the spin detection sensitivity to the point that it detects not the spin itself, but changes in the spin states [2].

Among the new spintronics active elements are surface-spintronic device [3], spintronic devices with constrained spintronic dopant [4], spintronic transistor [5], chemical spintronics sensor [10], carbon nanotube-based spintronic nanodevices [11] and others [6-9, 12]. Further parts of the lecture present important spintronic devices.

2. APPLICATIONS OF SPINTRONICS

2.1. Surface-Spintronics Device [3]

Now let us consider some of the most important spintronics devices invented until recently.

A surface-spintronic device operating on novel principles of operations may be implemented as a spin conducting, a spin switching or a spin memory device (Fig. (**1**)) [3]. It includes a magnetic atom thin film layered on a surface of a solid crystal and a drain and a source electrodes disposed at two locations on the magnetic atom thin film, respectively, whereby a spin splitting surface electronic state band formed in a system comprising said solid crystal surface and said magnetic atom thin film is utilized to obtain a spin polarized current flow. With electrons spin-polarized in a particular direction injected from the source electrode, controlling the direction of magnetization of the magnetic atom thin film allows switching on and off the conduction of such injected electrons there through. Also, with the use of the magnetization holding function of the magnetic atom thin film, it is possible to realize a spin memory device that can operate to write information on controlling the direction of magnetization of the magnetic atom thin film and that can operate to read information on detecting the state of conduction or nonconduction between the source and drain electrodes [3].

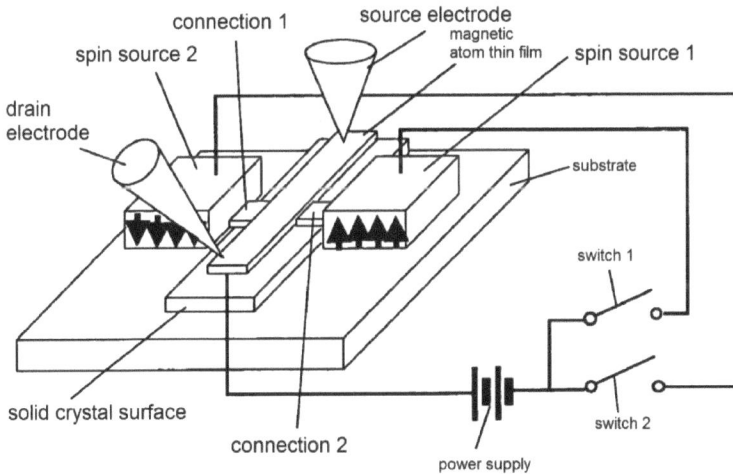

Figure 1: Schematic image of a surface-spintronics device [3].

How does it function? A surface-spintronic spin conducting device comprises a solid surface, a magnetic atom thin film layered on the solid surface, and electrodes mounted at two locations on the magnetic atom thin film, wherein a spin-splitting surface electronic state band formed in a system consisting of the solid crystal surface and the magnetic atom thin film is utilized to cause a spin current to flow. The solid surface is mostly a surface of nonmagnetic material having surface projected bulk band gaps, for example, a copper (111) surface or a covalent crystal surface so treated that it is terminated with hydrogen, and the magnetic atom thin film is a magnetic atom thin film having one to several atomic layers in thickness, *e.g.*, of iron atoms [3].

If electrons consisting only of up spin or electrons consisting only of down spin are supplied from the electrode of the spin conducting device, a spin current flows when the spin orientation of supplied electrons coincides with that of the surface electronic state band and no spin current flows when that is not the case. By controlling the direction of magnetization in the magnetic atom thin film, it is possible to make the spin orientation in the surface electronic state band coincident or not coincident with the spin orientation of the supplied electrons, and for this reason, it is possible to switch a spin current on and off and to realize a spin switching device. The surface electronic state band that can contribute to conduction can be made either of up spin only electronic state or down spin only electronic state, therefore it is possible to switch a spin current on and off at an efficiency of 100%. Also said spin conducting device can be used as a unit element for spintronic logic circuit and as a magnetoresistance element having an infinite changing rate of resistance. And also, it can also be used as a spin memory device, because a magnetization direction of the magnetic atom thin film, which is once controlled in one direction, remains held until next magnetization direction controlling is applied [3].

When the magnetization of an iron atom thin film is oriented upwards (then, the majority spin is the down spin ↓ and the minority spin is the up spin↑; this state is defined as a "normally polarized" state), the minority spin surface electronic bands **s1** and **s2** can be occupied with electrons exclusively of up spin. To wit, the electrons which can be injected into **s1** or **s2** and are allowed to propagate through the surface are electrons exclusively of up spin. On the other hand, when the magnetization of an iron atom thin film is oriented downwards (then, the majority spin is the up spin and the minority spin is the down spin; this state is termed to a "reversely polarized" state), **s1** and **s2** can be occupied with electrons exclusively of down spin. To wit, the electrons which can be injected into **s1** or **s2** and are allowed to propagate through the surface are electrons exclusively of down spin. This can be utilized to pass either a stream of electrons of up spin or a stream of electrons of down spin selectively and thus to pass a flow of perfect spin polarized electrons, namely a spin current, through the surface [3].

Now, in order to form a spin splitting surface electronic state band as mentioned above, it is necessary to form a magnetic atom thin film without destructing the crystallographic structure of a nonmagnetic crystal surface having a surface projected bulk band gap. It has been reported [3] that depositing an Fe atom thin film on a copper (111) surface by the laser MBE method permits forming such an iron atom thin film on the copper (111) surface without destructing its crystallographic structure. As another nonmagnetic crystal having a surface projected bulk band gap, there is Si (silicon) single crystal. With the recognition, however, that iron atoms tend to strip silicon atoms on a surface of the Si single crystal which is a covalent crystal, thereby forming suicide, it has hitherto been believed to be difficult to form an iron atom thin film on such a surface without destructing its crystal structure [3]. The device proposed by the authors [3] is computationally modeled. In particular, it is shown that if iron atoms are allowed to deposit on a Si (001) surface that has been terminated with hydrogen, it is then possible to form an iron atom thin film without destroying the Si (001) surface [3].

Fig. (**2**) shows computer experimental results, represented in crystallographic structural cross-sectional views, for surface structures of (a) a Si (001) surface that has adsorbed Fe atoms and (b) a hydrogen terminated Si (001) surface that has adsorbed Fe atoms. In Fig. (**2a**) it is seen that a Fe atom deforms the arrangement of surface Si atoms, then combining with a Si atom and that having iron atoms adsorbed to a Si (001) surface as it is destructs the Si surface crystal structure. On the other hand, it is seen from Fig. (**2b**) that the dimer structure of surface Si atoms is preserved and that having iron atoms adsorbed on a Si (001) surface that has been ended with hydrogen causes iron atoms to bond with Si without destructing the crystal structure of the surface. It follows, therefore, that a Si (001) surface terminated by hydrogen can be used as a nonmagnetic crystal surface having a surface projection gap for a spintronic device according to the invention [3].

In particular, it is important that the Si (001) surface is the major surface of a Si wafer for fabricating an integrated circuit in the current electronics, the ability to build a spintronic device according to the invention [3] on the Si crystal surface is advantageous, while t it facilitates hybridizing conventional electronic circuits with spintronic circuits [3].

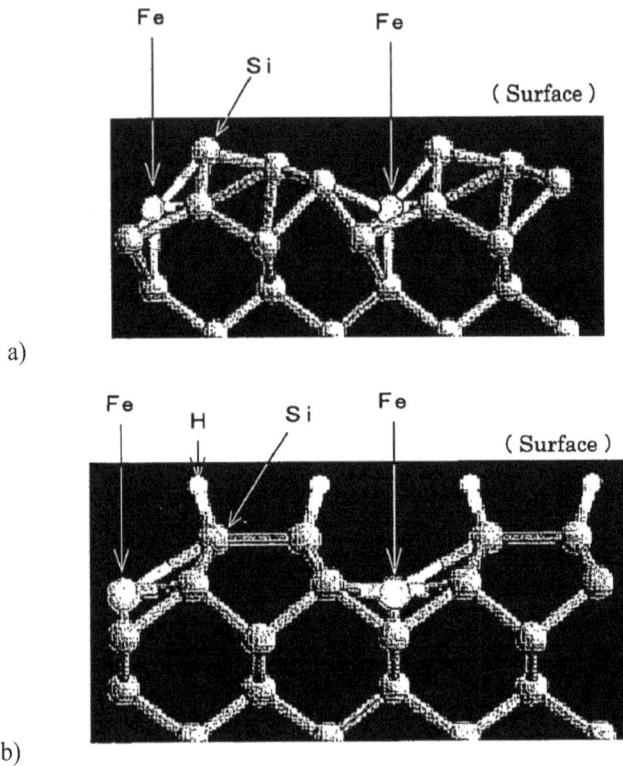

Figure 2: A graph illustrating computer experimental results for an electronic state band structure of a system of Cu (111) (a) surface having a Fe atomic layer deposited by the laser assisted MBE technology (b) [3].

The drain and source electrodes (Fig. (**1**) [3]) are mounted at two locations, respectively, on the magnetic atom thin film. Although the electrodes are each illustratively shown in the form of a probe for a scanning tunneling microscope (STM) for contact with the magnetic atom thin film, the contact may be by way of tunneling contact as in the ordinary use of STM as shown, namely by bringing the probe near the magnetic atom thin film surface to bring into point contact therewith, or otherwise by the usual way of sticking each electrode to the surface to establish facial contact therewith. By applying a bias voltage corresponding in energy to a surface electronic state band between the magnetic atom thin film and the source electrode, it is possible to inject from the source electrode into the thin film those electrons selectively, whose spin is identical in orientation to the spin of electrons in the surface electronic state band. Electrons so injected are taken out at the drain electrode which is higher in electric potential than the source electrode [3]. In this way, the electrons whose spin is identical in orientation to the spin of the surface electronic state band are caused to flow from the source electrode through the magnetic atom thin film to the drain electrode [3].

Returning to Fig. (**1**) [3], we can see that the magnetic polarity switching comprises the following building blocks: 1) two spin sources 1 and 2 which are magnetized parallel to an easy axis of magnetization of the magnetic atom thin film and mutually opposite direction; 2) two connections 1 and 2 that connect the magnetic atom thin film to the two spin sources 1 and 2, respectively; 3) a power supply for providing a bias voltage for spin injection; and 4) two switches 1 and 2 for it [3]. The spin sources 1 and 2 are made of ferromagnetic metals which are magnetized in the same directions which are identical to the directions of magnetization in which the magnetic atom thin film are to be normally and reversely magnetized, respectively. Advantageously but not exclusively, the spin sources 1 and 2 are the ferromagnetic metals magnetized upwards and downwards directed perpendicular to their surfaces, respectively, when the solid surface is a copper (111) surface and the magnetic atom thin layer is an iron thin film [3].

Further, for spin injection by applying a bias voltage between the spin source 1 or 2 and the magnetic atom thin layer, the spin sources 1 and 2 are connected to the magnetic atom thin film *via* the connections 1 and 2, respectively. The connections 1 and 2 may be sufficient if they permit spin injection from the spin sources 1 and 2 into the magnetic atom thin film, but are preferably made of nonmagnetic and electrically conductive material whose lattice constant is close to those of the atomic thin film and the spin sources 1 and 2 [3].

The switch 1 is a switch for normally polarized magnetization that can be turned on to apply a bias voltage of a selected magnitude from the power supply between the spin source 1 and the magnetic atom thin film [3]. And, the switch 2 is a switch for reversely polarized magnetization that can be turned on to apply a bias voltage of a selected magnitude between the spin source 2 and the magnetic atom thin film [3].

In the surface-spintronic spin switching device shown in Fig. (**1**) [3] with the second magnetic polarity switching means constructed as mentioned above, turning the switch 1 on causes the bias voltage to be applied between the spin source 1 and the magnetic atom thin film and normally polarized spins to be injected into the magnetic atom thin film from the spin source 1, thereby magnetizing the magnetic atom thin film in the normal direction. Thereafter, even with the switch 1 turned off, the magnetic atom thin film by its magnetization holding property remains in the state of magnetization in the normal direction [3].

Therefore, once the switch 1 is turned on, only electrons of up spin can propagate from the source electrode to the drain electrode through the surface electronic state band of the magnetic atom thin film [3]. The surface-spintronic spin switching device is rendered conductive when only electrons of up spin are supplied from the source electrode. The surface-spintronic spin switching device is rendered nonconductive when only electrons of down spin are supplied from the source electrode [3].

After the switch 1 is turned off, if the switch 2 is turned on, the bias voltage is applied between the spin source and the magnetic atom thin film to inject reversely polarized spins into the magnetic atom thin film from the spin source. This causes the magnetic atom thin film to be magnetized in the reverse direction. Thereafter, even with the switch 2 turned off, the magnetic atom thin film by its magnetization holding property remains in the state of magnetization in the reverse direction [3].

Therefore, once the switch 2 is turned on, only electrons of down spin can flow from the source electrode to the drain electrode through the surface electronic state band on the magnetic atom thin film. The surface-spintronic spin switching device is rendered conductive when only electrons of down spin are supplied from the source electrode. The surface-spintronic spin switching device is rendered nonconductive when only electrons of up spin are supplied from the source electrode [3].

After the switch 2 is turned off, if the switch 1 is again turned on, the magnetization of the magnetic atom thin film is switched again into the normal polarity direction so that the surface-spintronic spin switching device can conduct only electrons of up spin. Thus, supplied only with electrons of up spin from the source electrode, the surface-spintronic spin switching device is rendered conductive. Supplied only with electrons of down spin from the source electrode, the surface-spintronic spin switching device is rendered nonconductive [3].

In this way, the surface-spintronic spin switching device proposed by authors [3] functions as a spin switching device which is caused to switch its conductive and nonconductive states for a spin current when the polarity of its magnetization is switched by normally and reversely polarizing spin injections effected by the control means described above [3].

One of the most important challenges of the spintronics is a problem of room-temperature operating active elements. The solution is principally based on proper doping of semiconductor parts of the structures.

2.2. Spintronic Devices with Constrained Spintronic Dopant [4]

Various spintronic devices are based on diluted magnetic semiconductors (DMS). This family of semiconducting materials provides magnetic properties which are necessary for designing various active elements using spin injection. As noted in [4], current spintronics technology is limited by the currently used materials. It is important to have efficient spin carrier injection. It is also desirable to have manufacturing and operational compatibility with existing semiconductor processing technology. It is also desirable that the magnetic ordering or Curie temperature by at or near room temperature, instead of the more typical 100-200° K [4].

Such a spintronic device may include a superlattice and electrical contact coupled thereto [4]. The superlattice includes a number of groups of layers. Each group of layers may include a set of stacked base semiconductor monolayers defining a base semiconductor portion having a crystal lattice, one non-semiconductor monolayer constrained within the crystal lattice of adjacent base semiconductor portions, and a spintronic dopant [4]. The spintronic dopant may be constrained within the crystal lattice of the base semiconductor portion by the non-semiconductor monolayer [4].

The spintronic dopant may consist of at least one spintronic dopant monolayer adjacent the at least one non-semiconductor monolayer. This may be so, for example, where the energy levels favor attraction and retention of the spintronic dopant to the non-semiconductor. The spintronic dopant may comprise a transition metal, such as Manganese, Iron, and Chromium. Alternatively or additionally the spintronic dopant may also be a rare earth, such as a rare earth lanthanide, for example [4].

The non-semiconductor may be presented by Oxygen, Nitrogen, Fluorine, Carbon-Oxygen, and Sulphur. The semiconductor is to be Silicon, or more generally, may comprise a semiconductor selected from the group comprising Group IV semiconductors, Group III-V semiconductors, and Group II-VI semiconductors. The specific materials and structural configurations may be preferably selected so that the superlattice exhibits a Curie temperature of at least as high as room temperature [4].

The spintronic device proposed by authors [4] may be a spintronic field effect transistor (FET). Accordingly, the spintronic FET may include a substrate carrying a pair of superlattices in spaced apart relation to define a source and a drain, with a channel between the source and drain, and a gate adjacent the channel. Another type of the spintronic device is a spin valve. The spin valve may also include a substrate carrying a pair of superlattices in spaced apart relation with a spacer between the pair of superlattices [4].

Fig. (**3**) [4] plots an example of a spintronic device in the form of a spintronic field effect transistor (FET). The spintronic FET includes a semiconductor substrate carrying a pair of superlattices in spaced apart relation to define a source and a drain, with a channel between the source and drain, and a gate adjacent the channel. The gate includes a dielectric layer and a gate electrode [4].

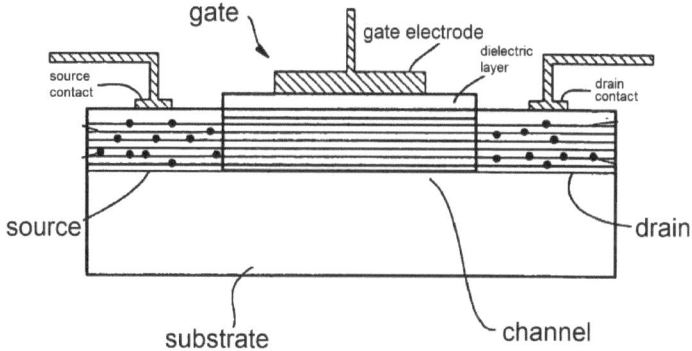

Figure 3: An example of a spintronic device in the form of a spintronic field effect transistor (FET) [4].

2.3. Spintronic Transistor [5]

Difficulties of semiconductor device fabrication are a limiting factor in the realization of Moore's law. Recently, the small size of conventional MOSFETs has created operational problems. Quantization of charge and energy levels become important in all materials at sizes below 10 nm. Furthermore, with small MOSFET geometries, the voltage that can be applied to the gate must be reduced to maintain reliability. To maintain performance, the threshold voltage of the MOSFET must be reduced as well. As threshold voltage is reduced, the transistor cannot be completely turned off, resulting in a weak-inversion layer which consumes power in the form of subthreshold leakage when the transistor should not be conducting. Subthreshold leakage can consume upwards of half of the total power consumption of a chip [5].

Various solutions are being developed to continue CMOS scaling beyond physical gate lengths of 20 nm (45 nm technology node). These solutions include non-classical CMOS architectures such as multiple gate and ultra-thin body MOSFETs, as well as different technology boosters such as mobility-enhancing strained Si, elevated source/drain, high-k gate dielectrics, and metal gate electrodes [5].

However, when the technology node reaches a length of about 22 nm (9 nm physical gate length), more radical innovation will be required. It had forecasted that CMOS architecture will continue to be used as a technology platform. Thus, there is a need for novel devices that significantly enhance system performance beyond that attainable with CMOS alone [5].

Spintronic devices, which utilize the intrinsic electron spin, are prime candidates towards this goal and can open routes toward combined storage and logic, memory transistors, reconfigurable logic, zero standby power technologies. Furthermore, devices using electron spin contribute toward long-term prospects of quantum computing and quantum cryptography [5].

As semiconductors become smaller, the classical laws of electricity and magnetism begin to break down, and the laws of quantum mechanics begin to govern the interactions within the semiconductor. The rules of quantum mechanics are quite different from the classical rules that determine the properties of conventional logic gates. If computers are to become smaller in the future, new quantum technology must replace or supplement conventional technology. It has been recognized that spintronics may be a route to solid-state quantum computing [5].

In addition to their mass and electric charge, electrons have an intrinsic quantity of angular momentum referred to as spin. Associated with the spin is a magnetic field, like that of a tiny bar magnet lined up with the spin axis. The two states of an electron are referred to as "spin-up" and "spin-down." In the presence of a magnetic field, electrons with spin-up and spin-down have different energies. In a conventional electronic circuit, the spins of the electrons are oriented at random and have no effect on current flow. A spintronic device will create spin-polarized currents and use the spin to store or process information [5].

Conventional views of spin-dependent transport in non-magnetic semiconductors have mainly focused on III-V semiconductors, where the direct band-gap allows convenient optical experiments to access (*i.e.*, read and write) the spin polarization of the carriers in the semiconductor [5].

Theoretical proposals for three terminal spintronic semiconductor devices have been made. A theoretical proposal for such a device assumes a half-metallic source/drain, wherein the highest occupied band is completely spin-polarized.

However, there are challenges to fabricate room-temperature half-metallic materials yet. Thus, the authors [5] have recognized a need to move toward Si based spintronic devices, where optical experiments are no longer possible [5].

The semiconductor device proposed by authors [5] includes: a substrate comprising silicon; a channel region formed on the substrate; a spin injector formed on the substrate at a first side of the channel region and configured to diffuse a spin-polarized current into the channel region; a spin detector formed on the substrate at a second side of the channel region and configured to receive said spin polarized current from the channel region; and a gate formed on the substrate in an area of this channel region [5].

The spin injector includes: a ferromagnetic material, and a dielectric material interposed between the ferromagnetic material and the substrate [5].

An all-electrical three-terminal device structure made with Si technology (and augmentations thereof, *e.g.*, strained Si on relaxed SiGe) and which can function as both a regular field effect (*i.e.*, charge) transistor and as a spin transistor is discussed by authors [5]. By simultaneously switching ON for both charge and spin flow when the gate bias is high and the source and drain magnetizations are parallel, and switching OFF when the gate voltage is low and/or the magnetizations are anti-parallel, the device can function as both a regular field effect transistor and a spin transistor [5].

There are three major prerequisites for successful spintronics implementation in semiconductors. First, there must be robust spin polarization/spin injection. Second, there must be efficient transfer across semiconductor interfaces. Third, there must be long lived spin-coherence [5].

Fig. (**4**) plots a spintronic transistor discussed in [5]. This spintronic transistor is an all-electrical spin transport device structure based on Si/SiGe. Silicon is an industry standard semiconductor. Si is a light element with small spin-orbit interaction (*i.e.*, a small shift in energy level due to an effective magnetic field). Thus, Si has very long electron spin lifetimes. Si is an ideal material for applications based on long spin coherence times, such as transistors with incorporated memory functionality, or quantum computing/cryptography applications [5].

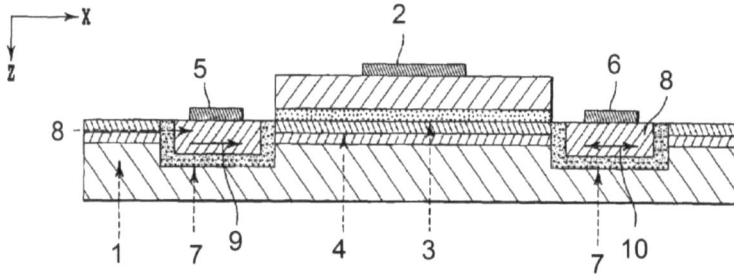

Figure 4: Spintronic transistor: 1- substrate, 2 – gate electrode, 3 – dielectric layer, 4 – channel, 5 – spin injector (source), 6 – spin detector (drain), 7 – dielectric, 8 – ferromagnetic (FM) metal, 9 – magnetization (direction), 10 – switchable magnetization [5].

The device shown in Fig. (**4**) includes substrate 1. Substrate 1 may include: silicon (Si); partially-depleted silicon-on-insulator (PDSOI); fully-depleted silicon-on-insulator (FDSOI); or virtual (relaxed) silicon-germanium (SiGe) substrate (on silicon or SOI). Exemplary embodiments of the present invention may also include a substrate including a layer of Si (depleted n^{++}-Si and n^--Si) and a layer of SiGe. The gate electrode 2 may be either a poly-silicon gate electrode material or a metal gate electrode material. The gate stack also includes dielectric layer 3 (either silicon dioxide dielectric material, silicon oxy-nitride dielectric material, or high-k dielectric material) [5].

The channel 4 is a degenerately doped channel that is fabricated by either: retrograde doping or hetero-layer growth (*e.g.*, strained silicon on a SiGe virtual substrate). The channel 4 is designed to be a heterostructure channel (strained silicon on a virtual SiGe substrate). This design was validated by Medici™ simulation. The following are the parameters of the Medici™ simulation that define the structure in the z-direction [5]:

(i)   gate work function is 4.8 eV;

(ii) electrical thickness of dielectric layer 3 is 15 angstroms (*i.e.*, a specific material was not selected; the physical thickness would be material dependent and chosen to give the same capacitance as 15 angstroms of SiO$_2$);

(iii) 10 nm relaxed Si$_{1-x}$Ge$_x$ (undoped);

(iv) 10 nm strained Si (undoped);

(v) 50 nm relaxed Si$_{1-x}$Ge$_x$ (N-doped to 10^{17} cm^{-3});

(vi) 1000 nm relaxed Si$_{1-x}$Ge$_x$ (N-doped to 10^{16} cm^{-3}); and

(vii) bulk Si (N-doped to 10^{16} cm^{-3}).

Layers (v) and (vi) are selected to be thick enough so that they relax in spite of the lattice mismatch with the bulk. Thus, they form a Si$_{1-x}$Ge$_x$ substrate thereby straining the thin Si layer (iv) on top. The strained layer forms a quantum well channel, and layer (vi) is used to modulation-dope the strained channel, *i.e.*, layer (iv). This helps to have a conductive channel while at the same time limiting spin relaxation due to ionized-impurity scattering [5].

In the above-described simulation x=0.4. Given a value for x, Medici™ can calculate Si$_{1-x}$Ge$_x$ band structure parameters internally. The band structure parameters (band gap and electron affinity) for the strained silicon layer atop of the relaxed SiGe are not calculated by the software. The band structure parameters are calculated by: (a) obtaining the electron affinity for Si$_{1-x}$Ge$_x$ using Vegard's Law (*i.e.*, liner interpolation between Si and Ge values), and (b) determining the conduction and valence band offsets for a layer of strained Si$_{1-y}$Ge$_y$ (y=0 in this case) on a layer of relaxed Si$_{1-y}$Ge$_y$. Conditions (a) and (b) immediately provide the aforementioned band structure parameters for the strained layer [5].

As shown in Fig. (**5b**) [5], when the gate bias (V$_g$) is zero, the channel region exhibits poor conductivity. As shown in Fig. (**5b**), the channel region (at about 0.02 μm) has an electron concentration that peaks just over 10^{16} cm^{-3}. As shown in Fig. (**6b**) [5], when the gate bias is 0.5 V, the channel region exhibits a much higher conductivity than when the gate voltage is zero volts. As shown in Fig. (**6b**), the channel region (at about 0.02 microns) has an electron concentration that peaks just over 10^{18} cm^{-3} [5].

a)

b)

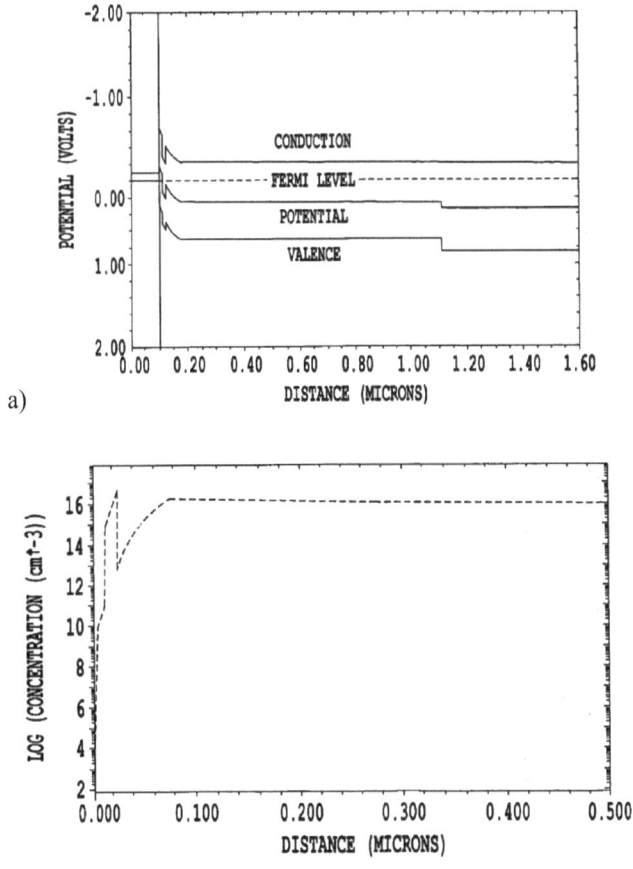

Figure 5: Energy band diagram of the strained silicon (a) and the electron concentration in the channel at $V_{gs} = 0$ (b) [5].

The channel based on a buried SiGe/Si heterojunction serves as both a confining layer to define the channel, and as a source of strain to lift the degeneracy of the six equivalent X-minima in the Si conduction band and influence the spin dynamics of the injected electrons. The reduced scattering in a strained channel would result in higher mobility and suppress the spin relaxation [5].

For spin transport in the SiGe layer, a very high Ge content (*e.g.*, 60-90%), as well as the implementation of spin injection contacts directly on the SiGe rather than on the Si is preferred. Ge concentration and strained SiGe/unstrained Si *vs.* unstrained SiGe/strained Si are exemplary parameters that can be used to influence spin transport [5].

a)

b)

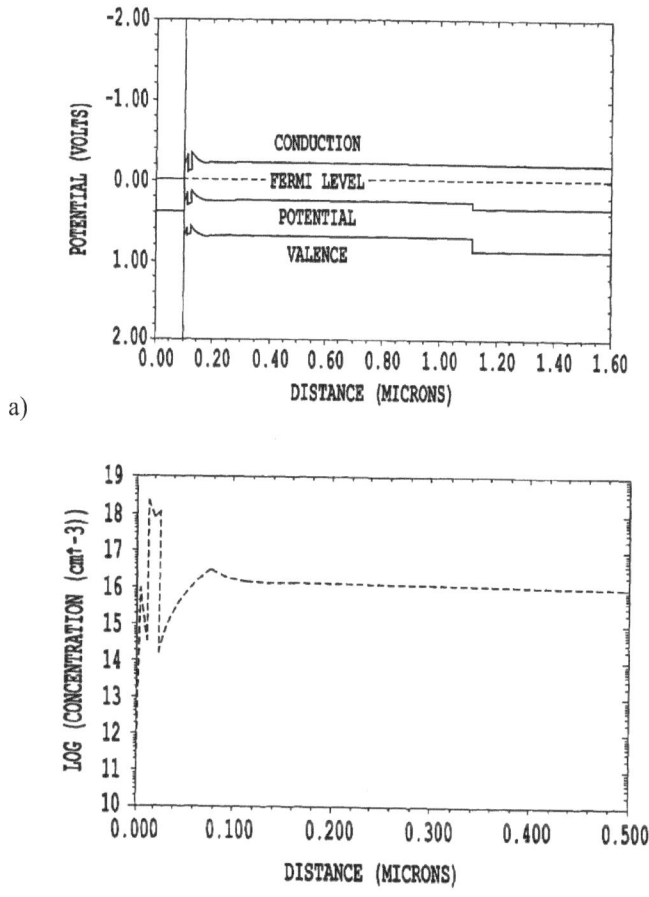

Figure 6: Energy band diagram of the strained silicon (a) and the electron concentration in the channel at $V_{gs} = 0.5$ V (b) [5].

The channel region is a quantum well channel. It is in this narrow region (in terms of depth along the z-direction (shown in Fig. (**4**)) that the source/drain tunnel barriers (described below) are sufficiently lowered by the application of the gate voltage. Obvious, that as the gate voltage is increased, the tunnel barrier is lowered and the concentration of spin polarized electrons in the quantum well channel increases. However, the concentration of charge carriers outside of the quantum well channel does not increase when the gate bias is applied. This is because, elsewhere along the z-direction, the semiconductor region has much higher barriers. These higher barriers provide less leakage current than conventional metal source/drain MOSFET designs [5].

The barrier formed by dielectric 7 is almost independent of the gate bias. It is the additional thermionic barrier that differs with the gate bias. The total tunneling barrier is a conjunction of dielectric 7 and the thermionic barrier. In the quantum well channel, the thermionic barrier in the x-direction (as defined in Fig. (**4**)) is also dependent on the position along the **z**-axis; specifically, whether the value of **z** is within the channel region or outside of the channel region. If outside the channel region, the thermionic barrier will vary somewhat, but is high for all gate voltages. It is only within a small range of z-values corresponding to the channel region that the thermionic barrier becomes relatively small for a high gate bias [5].

The spintronic transistor of Fig. (**4**) includes spin injector (source) 5 and spin detector (drain) 6. The spin injector includes FM metal 8 and dielectric 7. The spin detector also includes FM (ferromagnetic material) 8 and dielectric 7. Dielectric 7 forms a tunnel barrier between the FM 8 and the semiconductor material forming channel 4. The FM is a ferromagnetic metal or a ferromagnetic semiconductor material (for example, Co). Co and Co compounds may be deposited by conventional deposition techniques. The FM 8 in the spin injector has a magnetization 9 that is fixed in one direction. The FM 8 in the spin detector includes a magnetization 10 with a direction capable of being switched as explained below [5].

As shown in Fig. (**4**), spin injector 5 and spin detector 6 are each separated from the substrate and channel by a thin barrier of dielectric material 7 (for example, silicon dioxide, silicon oxy-nitride, or a high-k material) [5].

Ferromagnetic metal (FM) 8 and material 7 are selected so that electrons at the majority-spin Fermi level in the FM source encounter a low thermionic barrier to carrier injection in the ON state. As shown in Fig. (**6a**) [5], a source/drain material with a work function of 0.1 eV more than the electron affinity of silicon would result in a negligible thermionic barrier at the high gate bias. This, combined with the high channel conductivity, can lead to a high source to drain charge current (assuming that the magnetization of the source and drain are aligned). Because of band bending, a low gate voltage would result in a larger thermionic barrier for the same material (see Fig. (**5a**)). Thus, in conjunction with the poor channel conductivity, leads to a low source to drain charge current [5].

The spin injector and spin detector are formed from a layer of depleted n^{++}-Si formed on the substrate, an SiO_2 dielectric layer formed on the depleted n^{++}-Si layer, and a Co layer (*i.e.*, a ferromagnetic layer) formed on the SiO_2 layer [5].

Fig. (**7**) [5] shows schematic band diagram along the *x*-axis between the source and the drain of the semiconductor device of Fig. (**4**). As shown in Fig. (**7**), when the gate bias is low, a large potential barrier exists between the source and drain. Φ indicates the thermionic barrier height at the interface [5].

In contrast to Figs. (**7**) and (**8**) shows a schematic band diagram along the x-axis between the source and the drain when the gate is at a high bias. As shown in Fig. (**8**), the potential barrier between the source and drain is smaller and the thermionic barrier is negligible (as compared to Fig. (**7**)), which results in a higher charge carrier density in the channel region [5].

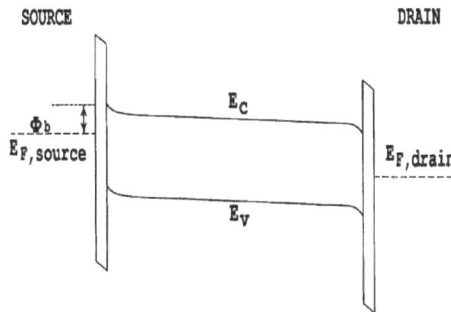

Figure 7: Schematic band diagram along the *x*-axis between the source and the drain of the semiconductor device of Fig. (**4**) [5].

The FM source/drain regions of the device described in [5] are contrary to conventional MOS transistors. Conventional MOS transistors have doped semiconductor source/drain regions that are doped opposite to that of the substrate or body (*e.g.*, *n*-type source/drain regions if the body is *p*-type material). A reversed biased *p-n*-junction, formed in the conventional MOS transistor, results in a low source drain current when the transistor is in the OFF state (*i.e.*, a low gate bias) [5].

As shown in Fig. (**4**) [5], the FM source/drain regions may be formed by etching a trench in the substrate and filling the substrate in with the FM. This effectively replaces conventional semiconductor source/drain regions. Alternatively, the metal source/drain regions may be raised as are more commonly seen in conventional fully-depleted silicon-on-insulator (FDSOI) devices. An exemplary transistor, with raised metal source/drain regions, is shown in Fig. (**9**) [5]. Fig. (**9**) shows a spintronic transistor that includes the following stack: $Co/SiO_2/depleted$ n^{++}-Si/n^--Si/strained SiGe, wherein the edge of the depletion layer coincides with the metallurgical junction between the n^{++}-region 11 and n^--region 12 [5].

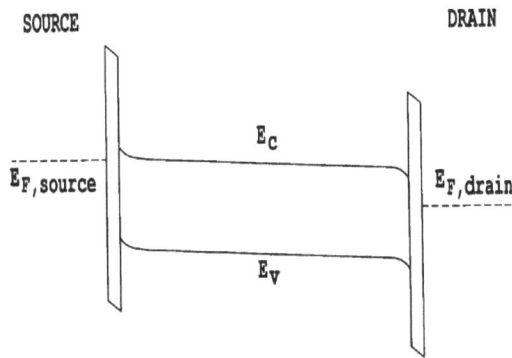

Figure 8: Schematic band diagram along the x-axis between the source and the drain of the device shown in Fig. (**4**) when the gate is at a high bias [5].

The transistor in Fig. (**4**) includes two magnetic tunnel contacts. The first magnetic tunnel contact is the spin injector 5, and the second magnetic tunnel contact is the spin detector 6. Magnetic tunnel contacts are a robust and reliable way towards the electrical injection and detection of spin-polarized electrons. The spin-dependent density of states in ferromagnetic metals or ferromagnetic semiconductors provides the necessary spin selectivity up to high temperatures. In a magnetized ferromagnetic material, the energy of one spin polarization is shifted higher with respect to the other. A current resulting from the magnetized ferromagnetic material is spin polarized because more electrons near the Fermi level are in one spin state than the other. For example, the Curie temperatures of Fe and Co are 770°C and 1130°C, respectively. The Curie temperature corresponding to a ferromagnetic material is the temperature below which the material is ferromagnetic. Above the Curie temperature, the magnetic ordering vanishes. Thus, the Curie temperature is the critical temperature of this phase transition [5].

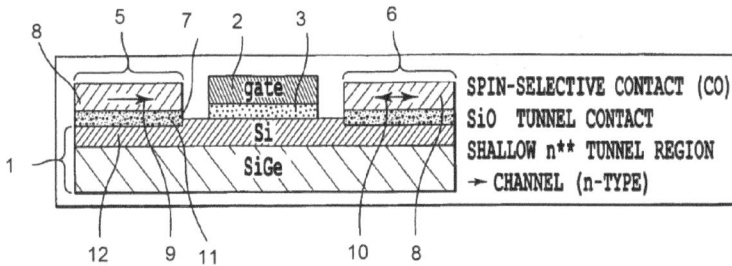

Figure 9: Spintronic transistor: 1- substrate, 2 – gate electrode, 3 – dielectric layer, 5 – spin injector (source), 6 – spin detector (drain), 7 – dielectric, 8 – ferromagnetic (FM) metal, 9 – magnetization (direction), 10 – switchable magnetization, 11 – shallow n^{++}-tunnel region, 12 – SiO tunnel contact [5].

The tunnel barriers allow for efficient spin injection from the metal into the semiconductor. The tunnel barrier is spin selective, *i.e.*, has different conductivities for up and down spins. [5].

There are metal (usually silicide) source/drain conventional MOSFETs where a Schottky barrier exists between the metal and semiconductor. However, a Schottky barrier (or Ohmic contact) would not inject a spin-polarized current. Thus, dielectric 7, shown in Fig. (**4**). replaces the Schottky barrier of conventional MOSFETs [5].

The tunnel barrier also provides the necessary matching in conductance between the magnetic contact and the semiconductor. A huge difference in conductivity between a ferromagnetic metal and a semiconductor precludes efficient spin injection from one into the other. Furthermore, the tunnel barrier forms a chemical barrier between the metal and the semiconductor, which improves the thermodynamic stability of the contact [5].

In the device proposed by authors [5] the tunnel barrier is formed by the ferromagnetic metal Co in combination with SiO_2 (as a dielectric 7). Co (with work function 5.0V) is a familiar material in conventional-CMOS technology (although mostly in the form of $CoSi_2$). Co is stable in contact with SiO_2 (*i.e.*, there is no tendency to decompose the SiO_2 and form either CoO or $CoSi_2$). When cobalt is placed in contact with silicon, the chemical potential in the materials will align with each other. The chemical potential in a semiconductor is doping-dependent. The band-bending at the interface is the difference in the work functions. The band-bending determines if, and how easily, carriers might be injected from one material into the other. For example, the band-bending might create a (Schottky) barrier that inhibits carrier injection [5].

Important considerations when designing the tunnel barriers are the electric quality of the tunnel barriers and the magnetic quality of the magnetic tunnel contacts (*e.g.*, the interface spin polarization of the *Co/oxide* interface, and the absence of spin-flip scattering at the oxide/semiconductor interface) [5].

Dangling bonds in Si have unpaired electrons, which might result in spin-flip scattering. Thus, in one exemplary embodiment of the present invention, SiO_2 /*Si* interfaces, which can essentially be passivated perfectly, are employed. The thicknesses of the SiO_2 /*Si* interfaces are scaled down to smaller thickness (*i.e.*, 8-12 angstroms) to allow tunneling [5].

High magnetoresistance in *ferromagnet/barrier/non-magnet/barrier/ferromagnet* structures is achieved when: r_N $(l_N/\lambda_N) \ll r_c \ll r_N \cdot (\lambda_N/l_N)$ (*i.e.*, efficient spin injection condition), where l_N is the length of the non-magnet (N) region (*i.e.* channel length) and λ_N is the spin diffusion length in the N-region, and $rN = \lambda N \sigma N$ is its effective resistance (σ_N being the conductivity of the N-region), and r_C is the tunneling resistance for the source/drain barriers. Considering, as a typical example, $l_N \sim 50$ nm and $\lambda_N \sim 1000$ nm, the above criterion translates to $0.05 r_N \ll r_c \ll 20 \cdot r_N$. Imposing an even stricter inequality provides the following rule for designing the source/drain tunnel barriers: $0.5 r_N < r_c < 2 r_N$. This inequality is satisfied for the ON state. However, this inequality is not satisfied for the OFF state. In the OFF state, $r_N^{off} \approx 100 \cdot r_N^{on}$ [5].

From Figs. (**5b**) and (**6b**) one can see that the ratio of the carrier density in the channel between the ON and OFF states is about 100. Therefore, σ_N should also be about two orders of magnitude larger in the ON state than in the OFF. Then the ON and OFF state channel resistances are related as follows: $r_N^{off} \approx 100 \cdot r_N^{on}$. The *source/drain* tunnel barrier thickness is chosen so that $0.5 r_N^{on} < r_C < 2 r_N^{on}$ and high magnetoresistance is realized in the ON state; the inequality is then not satisfied for r_N^{off}, and we do not have efficient spin injection in the OFF-state. Thus, the device would have a high magnetoresistance in the ON state (*i.e.*, dependence of the current on the relative orientation of the magnetizations in the source and drain) but not in the OFF state (*e.g.*, no dependence of the current on the relative orientation of the magnetizations in the source and drain). In the OFF state, the transistor passes a small leakage current with no appreciable magnetoresistance [5].

Several variables enter the calculation to determine how thick the tunnel barriers (*i.e.*, the dielectric 7) should be as follows: [5]:

- o   the spin-dependent chemical potentials in the ferromagnetic source;

- o   the barrier height (*i.e.*, the barrier can be assumed to be rectangular);

- o   that spin-up and spin-down carriers see the same type of barrier; (*i.e.*, both spin-up and spin-down carriers tunnel into propagating states in the semiconductor on the other side of the barrier);

- o   the electronic structure parameters of the materials, *e.g.*, the effective mass (*i.e.*, the effective mass can be assumed to be constant throughout);

- o   geometrical parameters.

Below, it is shown for realistic parameters, that the efficient spin injection condition can be satisfied. For this example, the condition will be strengthened to show a strict inequality for designing the tunnel barrier [5]:

$$10 \cdot r_N \cdot (l_N / \lambda_N) < r_c < 0.1 \cdot r_N \cdot (\lambda_N / l_N) \tag{2.1}$$

For a channel concentration in the ON state of $n \sim 10^{20}$ cm^{-3}, the conductivity is

$$\sigma_N{}^{on} \approx ne\mu = 10^{20} \cdot 1.6 \cdot 10^{-19} \, C \cdot 1000 \; cm^2 \, V^{-1} \, s^{-1} \approx 10^4 \, (\Omega \cdot cm)^{-1}, \tag{2.2}$$

where e is the charge of an electron and μ is the electron mobility. For channel length $l_N \sim 30$ nm and spin relaxation length $\lambda_N \sim 1000$ nm,

$$r_N{}^{on} = \lambda_N / \sigma_N \approx 10^{-8} \, \Omega \cdot cm^2 \tag{2.3}$$

Inequality (2.1) may be rewritten as:

$$0.3 \cdot r_N^{on} < r_c < 3.33 \cdot r_N^{on} \tag{2.4}$$

σ_N^{off} and r_N^{off} are significantly lower than the corresponding ON quantities, so that condition (2.4) is not satisfied in the OFF state [5].

The tunnel barrier contact conductance is defined as:

$$r_c = (\Sigma\uparrow + \Sigma\downarrow)/ 4 \Sigma\uparrow\Sigma\downarrow \tag{2.5}$$

wherein $\Sigma\uparrow$, $\Sigma\downarrow$ are the spin-up and spin-down contact conductance per unit area. From Landauer-Buttiker theory, the spin contact conductance per unit area is approximated by:

$$\Sigma = (e^2 m^*/h\pi\hbar^2)T(\varsigma)\cdot k_B T \tag{2.6}$$

where e is the electronic charge, h is the Planck constant, $\hbar = h/2\pi$, $T(\zeta)$ is the energy-dependent tunneling probability, $m^* \approx 0.5 m_0$ the effective mass, and $k_B T$ is the thermal energy range. It is assumed that the tunneling probability is constant within this thermal energy range below the chemical potential, and is negligible for lower energies [5].

Using the WKB approximation, one writes the tunneling probability as: $T \approx \exp(-2\kappa b)$, where:

$$\kappa(E) = (2m^*(V - E)\hbar^{-2})^{1/2} \tag{2.7}$$

and b is the barrier thickness. Authors [5] have chosen $V-E_\downarrow = 2$ eV and $V-E_\uparrow = 2.5$ eV (for comparison, a $Si/SiO_2/Si$ system with degenerate silicon gives $V-E = 3$ eV for electrons) to get $\kappa_\downarrow$ and $\kappa_\uparrow$. Substituting into equation (2.7), and assuming a barrier thickness $b \approx 1$ nm, one gets $T_\uparrow$ and $T_\downarrow$, and from equation (2.6) one obtains $\Sigma_\uparrow$ and $\Sigma_\downarrow$ [5].

Finally, equation (2.5) gives:

$$r_c \sim 10^{-8} \, \Omega \cdot cm^2 = r_N, \tag{2.8}$$

which satisfies the condition in inequality (4). Variation in the channel conductivity from the typical values used here can be easily accommodated, while still maintaining efficient spin injection, by slightly adjusting the barrier thickness since the barrier contact resistance depends exponentially on it [5].

Fig. (**10**) is a graph of the boundaries for the efficient spin injection inequality for the ON state and the OFF state. The graph in Fig. (**10**) is a plot of the lower and upper bounds for the ON state and OFF state obtained from inequality (2.1). The graph shows a plot of barrier resistance multiplied by area *vs.* channel length (l_N). The conductivity of the channel region is different in the ON state and the OFF state (*i.e.*, high and low gate biases respectfully). Inequality (2.1), written for both the ON state and OFF state results in:

$$10 \cdot [r_N^{on} \cdot (l_N / \lambda_N)] < r_c < 0.1 \cdot [r_N^{on} \cdot (\lambda_N / l_N)] \tag{2.9}$$

(for the ON state) and:

$$10 \cdot [r_N^{off} \cdot (l_N / \lambda_N)] < r_c < 0.1 \cdot [r_N^{off} \cdot (\lambda_N / l_N)] \tag{2.10}$$

(for the OFF state). Here r_c is chosen to satisfy inequality (2.9), but not inequality (2.10) [5].

Operation of the spintronic transistor proposed by authors [5] will be made with reference to Fig. (**4**). The device has source/drains, dielectric layer 3 is a high-k dielectric, the gate electrode is metal, dielectric layer 7 (*i.e.*, the source/drain injection barrier) is silicon dioxide or silicon-oxynitride, and the channel is a modulation-doped Si/SiGe based quantum well channel. To allow tunneling, dielectric layer 7 is specifically chosen not to be a high-k dielectric [5].

When zero voltage is applied to gate 2, the tunnel barrier is sufficiently thick to prevent electrons from ferromagnetic metal (FM) 8 from tunneling into channel region 4. Thus, when the gate bias is zero, channel 4 is designed to be OFF (*i.e.*, no spin-polarized current flows through the channel) when the gate bias is small. When the gate bias is sufficiently large, and the magnetization 9 and 10 of the drain and the source are parallel, the transistor is considered ON (*i.e.*, spin-polarized current flows from the source to the drain). The source/drain current (which is a spin-polarized current) is injected through tunnel barriers 7. When the gate bias voltage is high, but the magnetizations 9 and 10 of the source and drain are anti-parallel, the transistor is OFF (*i.e.*, no spin-polarized current flow through the channel) [5].

During operation of the spintronic transistor, a magnetization of the drain is parallel or anti-parallel to that of the source. This could be implemented as in an MRAM or by other magnetization switching techniques. In the OFF state, the transistor passes a small, *i.e.* 'leakage', charge current with no appreciable magnetoresistance (that is, no dependence of the current on the relative orientation of the magnetizations in the source and drain). In the ON state, the device passes a high spin-polarized current when the source and drain magnetizations are parallel, but it passes a low current when they are anti-parallel since this is also a high-magnetoresistance configuration. Thus, in the parallel configuration, a non-limiting embodiment of the device shown in Fig. (**4**) can be used as a "charge transistor." A non-limiting embodiment of the device shown in Fig. (**4**) can also provide non-volatile memory functionality as a "spin transistor" whose state can be read by measuring the source to drain current after switching the "spin transistor" ON [5].

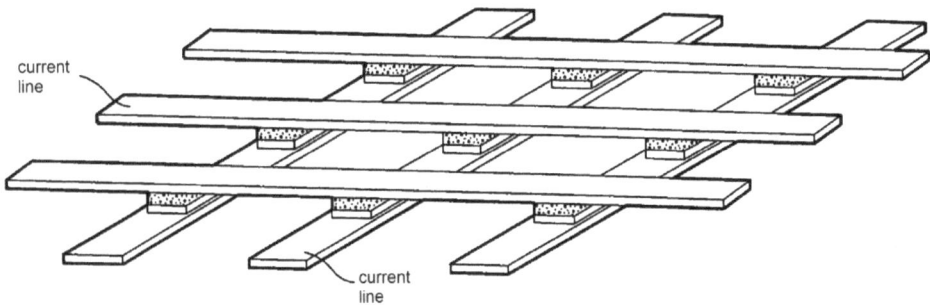

Figure 10: Schematic image of the structure of metal wires that switch the direction of the magnetization of the spin detector [5].

Fig. (**10**) shows an exemplary structure of metal wires that switch the direction of the magnetization of the spin detector. During a write operation, a current is forced along current lines [5].

The currents create a magnetic field, whose vector sum is sufficient to switch the direction of the magnetization of the FM in the spin detector. Thus, this spintronic transistor is capable of performing a logical operation (*i.e.*, ON/OFF) by utilizing the spin of the electrons [5].

Furthermore, a spintronic transistor presented by authors [5] is capable of being used as a non-volatile memory device. Each spintronic transistor can store one bit of information. Information can be stored in the unpinned ferromagnetic material of the drain even if the power is turned off. In addition, since the spintronic transistor does not need to be updated continuously, power consumption is much lower than conventional transistors [5].

Performing a function as a memory, the active element described in [5] can store either a *zero* or a *one*. To perform read operation on the spintronic transistor, a bias is applied to the gate. If a current is detected at the drain, then the magnetization of the source and drain are parallel, and this can represent a *1*. If no current is detected at the drain, then the magnetizations of the source and drain are anti-parallel, and this can represent a *zero* [5].

Consequently, spintronic devices can represent data with far fewer electrons than conventional charge-based electronics. Accordingly, spintronic devices are considerably smaller and faster than conventional microelectronics [5].

Furthermore, the function of spintronic transistor [5] could be changed "on the fly" by changing the magnetization of the FM in the drain (or even the source). Processors using a spintronic transistor invented by authors [5] could readjust mid-clock cycle by reversing the magnetization of some of its elements in order to do calculations more efficiently. This allows an entirely new approach to computing, which is software driven (rather than hardware-determined), and a standardized reprogrammable logic chip would become a universal microprocessor [5].

As described above, the magnetization of the spin detector (*i.e.*, the drain) is either parallel to or anti-parallel to the magnetization of the spin injector (*i.e.*, the source). The switching of the magnetization in the drain is implemented as in an

MRAM or by other magnetization switching techniques. For example, the switching is performed with a global static magnetic field that switches the drain magnetization but not that of the source. This is achieved by shaping the source/drain regions differently, so that one region switches magnetization at a lower magnetic field than the other region. In another exemplary embodiment of the present invention, the semiconductor device would include metal lines addressing the drain as in an MRAM. The magnetic field due to the current in these metal lines will switch the relative magnetization of the drain. The source magnetization is held fixed, for example, by using an anti-ferromagnetic pinning layer [5].

2.4. Nanodevices for Spintronics [10]

Progress in the miniaturization of traditional solid state electronics based on manipulating electronic charges is rapidly approaching the natural technological limit imposed by the discrete atomic structure of matter. Already working at the nanometer scale, current technology not only encounters significant challenges requiring ever-increasing nanoscience research and design effort, but after a more than six order-of-magnitude reduction in size of the electronic components achieved by the beginning of this century, there is less than an order of magnitude left to go. Consequently, there has been a growing recognition that one of the avenues for future progress rests with a new approach in electronics, dubbed spintronics, where not only electron's charge, but also its quantum spin degree of freedom is manipulated [10].

The first application of a spintronic effect, giant magnetoresistance or GMR, can be found in magnetic sensing devices. GMR technology uses one type of electron spin manipulation, the change in resistance of a resistive element in the presence of a magnetic field. GMR elements can therefore be considered as spintronic analogues of conventional resistors. These devices have been applied to magnetic read heads for computer hard drives and magnetic random access memory (MRAM). They also garnered the 2007 Nobel Prize in Physics for discoverers Albert Fert and Peter Grünberg [10].

A number of materials, including natural half-metals, such as chromium dioxide, doped perovskite manganites, and various magnetic semiconductors have been investigated as possible sources of spin-polarized electrons for spintronic devices [10].

At present, available devices sensitive to the polarization of electric current utilize the GMR-dependence of the resistance on the spin orientation in alternating ferromagnetic and antiferromagnetic or magnetic and nonmagnetic multilayers, *e.g.* in Iron/Chromium/Iron (Fe/Cr/Fe) trilayers. For a Cr interlayer of appropriate thickness, the coupling between the adjacent ferromagnetic iron layers is antiferromagnetic, and in the absence of an external magnetic field their magnetizations are antiparallel. An external magnetic field can co-align magnetization of magnetic layers and this decreases the amount of spin-dependent electron scattering, decreasing the resistance. At present, approximately 90% of all hard drive read heads use the GMR technology which allows the storage density to be increased by over two orders of magnitude compared to earlier technology [10].

Other approaches to producing spin-polarized electric currents are based on magnetically doped semiconductors, such as $(Ga_{1-x}Mn_x)As$, $(In_{1-x}Mn_x)As$, $(Be_{1-x}Mn_xZn_y)Se$, $Cd_{1-x}Mn_xTe$, *etc.*, where the properties of ferromagnetic and semiconducting systems are combined on the material level. Both semiconducting properties and carrier-controlled magnetism of these materials are mediated by the same doped charge carriers. Hence, there exists not only direct coupling between the two, but also an intricate interplay between magnetic and semiconducting properties, requiring a fine doping/compositional optimization, which has hindered technological progress so far. Similar material chemistry problems are creating obstacles on the path of using natural semimetals with spin-polarized electronic bands, such as Fe_3O_4, or $La_{0.7}Sr_{0.3}MnO_3$ [10].

Thus, realizing the complications associated with dependence on only compositional techniques to utilize spin-polarized bands/currents, the inventors propose a multi-layer approach, making use of the particularly attractive properties of graphene [10].

The authors of [10] present a spintronic device, an apparatus that includes a spintronic device, and methods of using and making spintronic device(s) that utilize a graphene sheet having first and second surfaces and at least two electrode terminals [10].

The graphene sheet can have a thickness less than about 10 nm, or can have a thickness of a single monolayer of carbon, or can have a thickness of multiple monolayers of graphene. In some embodiments, the graphene sheet can possess a non-zero magnetization. An electrical property of the graphene sheet is based on a magnetic field effect induced in the graphene sheet [10].

Two electrode terminals are electrically coupled to the graphene sheet and are spaced apart. The graphene sheet is configured to conduct an electric signal between the at least two electrode terminals and is configured to affect an electrical signal based on the magnetic field effect induced in the graphene sheet. The electrical properties of the graphene sheet are related to the magnetic field effect induced in the graphene sheet [10].

The spintronic device can also include one or more magnetic materials, where at least one of the magnetic materials is in contact with at least a portion of the first or second surface of the graphene sheet. The magnetic material can have a non-zero magnetization which induces the magnetic field effect in the graphene sheet. One of magnetic materials is disposed on the graphene sheet as a continuous layer, or as a plurality of non-continuous discrete sections. It can be further covered with properly mated additional magnetic layer(s). The magnetic materials can include ferromagnetic and antiferromagnetic materials. The ferromagnetic material can be in contact with the graphene sheet, or the antiferromagnetic material can be in contact with the graphene sheet and the ferromagnetic material can be in contact with the antiferromagnetic material, and combinations of order may be considered. Contacting the graphene sheet with a metallic layer, whether or not part of the device structure, may lead to poorer device performance. Magnetic moments of the magnetic materials can be correlated and the alignment of the magnetic moments of the magnetic materials can also be correlated [10].

To use a preferred embodiment of a spintronic device having a graphene magnet multilayer (GMM), an electrical input signal, such as an electric current, is supplied to a first electrode terminal that is in electrical contact with a graphene sheet. The electrical input signal passes through the graphene sheet, in the plane of the sheet, and the graphene sheet affect the electrical input signal based on a magnetic field effect induced in the graphene sheet. An electrical output signal is obtained at one or more of the at least two electrode terminals in response to the electrical input signal. The electrical output signal is based on the electrical input signal and the magnetic field effect induced in the graphene. A magnitude of the magnetic field effect can be varied to vary an electrical property of the graphene sheet. The electric input signal can be polarized based on the magnetic field effect induced in the graphene sheet. A magnetization in a magnetic material of the GMM can be induced, which in turn induces the magnetic field effect in the graphene sheet. Different magnetic field effects can be induced in different regions of the graphene [10].

To form a spintronic device a graphene sheet is disposed on a substrate which may be a non-magnetic or a first magnetic material. Additionally, at least two electrodes are disposed on the graphene sheet to provide regions for electrically coupling to the graphene. A magnetization is induced in the first magnetic material. In some structures proposed by authors [10], the first magnetic material can be an antiferromagnetic material and a second magnetic material of a ferromagnetic material can be formed on the first magnetic material [10].

Some structures proposed by the authors [10] can be used to create and control devices based on spin-polarization in the presence of a magnetic field effect induced in a graphene sheet, such as spin-controlled transistors and logic gates, and the like. Active, re-writable, and re-configurable devices can be formed in accordance with the preferred embodiments of the invention [10] so that the function of these devices can change depending on the magnetization pattern of an underlying magnetic material [10].

The active elements invented in [10] exploit spin-dependent splitting of the electronic bands in graphene based on a magnetic field effect induced in the graphene due to a magnetic proximity effect [10].

As authors [10] point out, "graphene" refers to a two-dimensional (2D) crystalline form of carbon. Graphite (a bulk form of carbon) and graphene form with hexagonal symmetry. Carbon nanotubes (CNT) may be thought of as graphene sheets rolled into tubes exhibiting one-dimensional (1D) properties [10].

An important practical requirement for spintronic devices is that they operate at room temperature. To accomplish this, spin polarization is obtained and maintained at temperatures where thermal energy $k_B T$ (k_B is the Boltzmann constant and T is temperature) equals or is less than the electron's magnetic Zeeman energy in a magnetic field that is typically greater than 220 Tesla, which is more than two orders of magnitude stronger than artificial fields available in magnetic recording devices [10].

One solution for implementing spintronic devices that operate at room temperature is to use magnetic half-metals or semiconductors, where the interplay of the Coulomb interaction and the Pauli exclusion principle, also known as the electron spin exchange interaction, results in a spin-dependent electronic band splitting such that the conduction band is polarized with respect to magnetization. Exchange is responsible for the ferromagnetism of common metals, such as iron and nickel, and the antiferromagnetism in their oxides. These magnetic properties can be maintained at temperatures up to several hundred degrees Celsius. Hence, magnetic band splitting caused by exchange can also be very large, and comparable to thermal energies at and above room temperature. While exchange can be viewed as an effective Weiss molecular (exchange) magnetic field acting on electrons in magnetic medium, this effective magnetic field originates from the electrostatic Coulomb interaction. For this reason, the exchange field can be much stronger than artificial magnetic fields from electromagnets, which are electrodynamic, caused by non-relativistic motion of electric charges, and therefore contain a "relativistic" small factor, $\alpha \approx 1/137$, compared with $\alpha \approx 1$ in GMM structures. Another feature of the Weiss exchange field is that it only couples to spin [10].

Fig. (**11a**) to (**11c**) illustrate the low-energy electronic dispersion in graphene 1 for applied magnetic fields (H) of differing strengths. As shown in Fig. (**11a**), the low-energy electronic structure in graphene 1 results from the sp^2 C—C bonding in the hexagonal carbon layer and consists of two delocalized

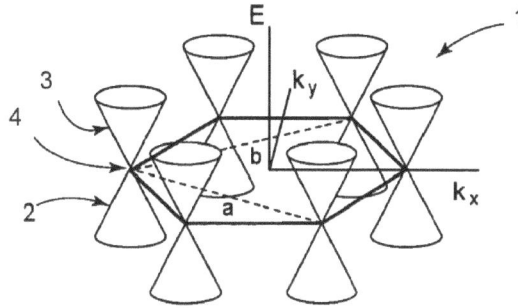

Figure 11a: [10]

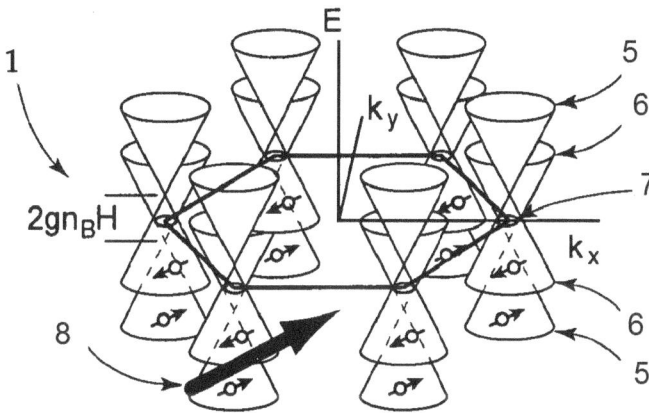

Figure 11b: [10]

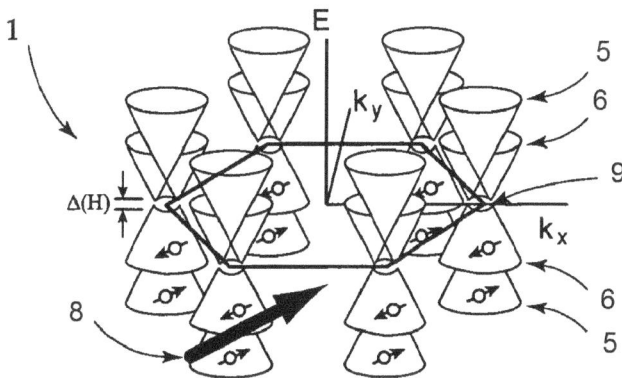

Figure 11c: Fig. (**11a**) through (**11c**) illustrate the low-energy electronic dispersion in graphene 1 for applied magnetic fields (H) of differing strengths.

valence bands, a filled (electron) "bonding" π-band 2 and an empty (hole) "antibonding" π^*-band 3, which meet at a single point at the corners of the Brillouin zone (BZ) 4. In an ideal half-filled case corresponding to undoped graphene in zero magnetic field (Fig. (**11a**)), this point lies exactly at the Fermi energy, resulting in a Fermi surface consisting of a set of points at BZ 4 corners. This results in a linear 2D dispersion, $\varepsilon(k)=v_f k$, where v_f is the fermion velocity, characteristic of massless fermions whose motion is governed by the 2D relativistic Dirac equation. Dotted lines show unit vectors a and b of the hexagonal lattice of a honeycomb graphene layer, containing two C atoms. Hence, there are two inequivalent Dirac points per hexagonal unit cell, forming the Fermi surface in neutral (undoped) graphene in the absence of magnetic field effect [10].

In Fig. (**11b**), the magnetic field (H) 8 parallel to the graphene 1 layer introduces Zeeman splitting of $2g\mu_B H$ between the parallel spin (P) band 5 and the antiparallel spin (AP) band 6, where g stands for the Landee factor and μ_B for the Bohr magneton. Hence, P and AP bands 5 and 6 acquire coincident, electron-type and hole-type Fermi-surfaces, respectively, whose radius 7 is $k_f = g\mu_B H/v_f$. For strong enough magnetic fields (H) 8, electron-electron interactions lead to an excitonic instability, opening a gap Δ(H) 9 in the electronic dispersion, as depicted in Fig. (**11c**) [10].

When graphene is mated to a magnetic (FM or AFM) material the graphene can be magnetized by the Weiss exchange field by virtue of the magnetic proximity effect to induce a magnetic field effect in the graphene. In this case, the completely polarized electron and hole bands of Fig. (**11b**) and (**11c**) may result [10].

Several technologies for producing well-ordered monolayer (MG) and few layer (FLG) graphene, *e.g.*, a graphene sheet, on different substrates can be implemented. In a first approach, depicted in Fig. (**12a**), high-quality MG and/or FLG 1 are grown either by chemical vapor deposition (CVD) or epitaxially in ultrahigh vacuum (UHV) on a hot reactive magnetic material substrate 2, such as metal or metal carbide (*e.g.*, Ni(111), TiC(111) surface), at temperatures approximately 500°-600°C and with carbon atoms supplied by the dissociation of a hydrocarbon gas such as ethylene or propylene, thus, producing GMM 3.

Dramatic reduction of the hydrocarbon dissociation rate, up to 10-100 times, upon the MG layer allows precise control of the number of graphene layers deposited in this process. Using this method, MG or FLG layers can be formed on an amorphous ferromagnetic alloy such as CoPtCrB, or on the resulting graphene sheet, in addition to metal and carbide substrates. After the graphene is deposited, the graphene sheet can be capped by another material if desired. The order of deposition depends on the materials employed and the ultimate use of the GMM device [10].

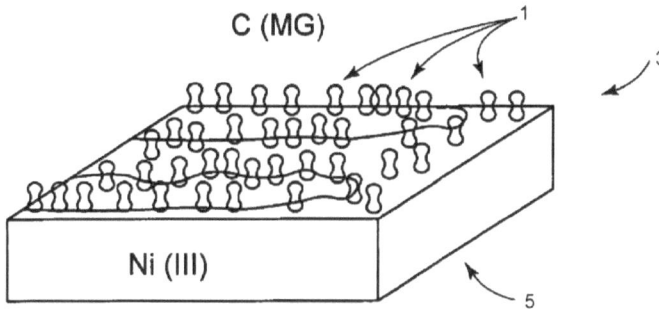

Figure 12a: [10]

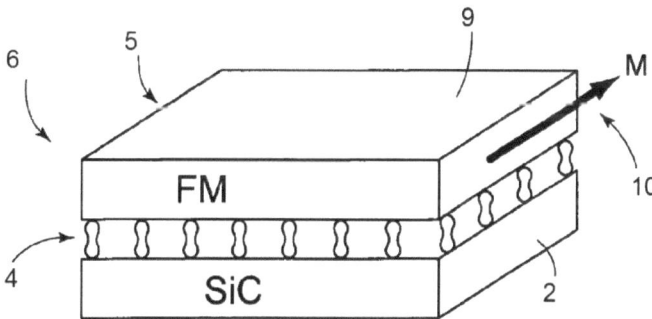

Figure 12b: [10]

In a second approach, shown schematically in Fig. (**12b**), controlled layer-bylayer growth of epitaxial graphene 1 can be achieved by sublimation of Si from the substrate **208** surface, such as a silicon carbide (SiC) substrate, during thermal annealing in ultrahigh vacuum (UHV) at 1200° to 1500°C. By controlling the exposure time (typically 1 to 20 minutes) and the temperature of the solid state evaporation, unconstrained high-quality heteroepitaxial monolayer; bi-layer and few-layer graphene 4, as well as millimeter-sized single-crystalline graphite, can be grown. In this method the substrate 2 can be SiC, a wide-gap (~3eV) semiconductor which is well suited for applications in semiconductor devices and commercially available in the form of high-quality oriented single-crystalline wafers. MG and FLG sheets obtained in this way have been shown to possess long-range structural coherence with domain sizes of 90 nm or more, high electron mobility with an elastic mean free path of about 600 nm, phase coherence lengths exceeding 1 μm at 4 K, and the ability to sustain high current densities of about 1 nA per carbon atom, similar to carbon nanotubes [10].

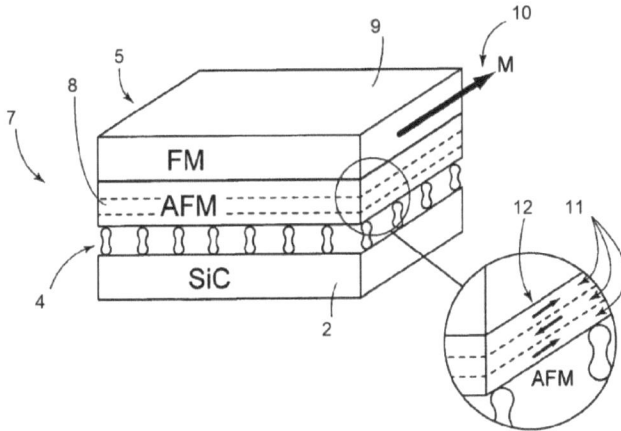

Figure 12c: [10]

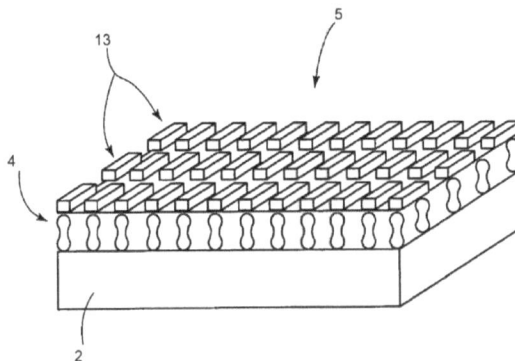

Figure 12d: Fig. 12 (**a-d**) shows technological processes for producing the device described in the text.

A popular technique for obtaining flakes of graphene is to "peel off" layers from a graphite source. In this case, the graphene may be temporarily affixed to a substrate while (anti-)ferromagnetic layers are formed [10].

As plotted in Fig. (**12b**) and (**12c**), resulting layers of graphene 1 (*e.g.* a graphene sheet) formed on SiC can be capped by depositing magnetic material(s) 5 of choice at a lower temperature, down to room temperature, thus forming graphene-(antiferro)-ferro magnet multilayers 6 and 7 as seen in Fig. (**12b**) and (**12c**), respectively. In Fig. (**12b**), a graphene sheet 4 (MG or FLG) is topped with a layer of ferromagnetic (FM) material 9 such as Ni, CoPtCrB, or other FM material. Fig. (**12c**) depicts an embodiment in which the graphene sheet 4 is topped by an AFM material 8, such as NiO or CoO, which in turn is topped by a FM material 9 [10].

In ferromagnetic (FM) materials, electron spins are aligned parallel to an applied magnetic field leaving a spontaneous magnetization even after an external magnetic field, the inducing field, is removed. In antiferromagnetic (AFM) materials, adjacent spins are aligned: one parallel and one anti-parallel, so they show little or no spontaneous magnetism. Ferrimagnetic materials have two different kinds of magnetic ions; adjacent ions alternate spins parallel and antiparallel to an applied field, but not equally, resulting in a net magnetization [10].

Still considering Fig. (**12a**) to (**12c**), in structures proposed by authors [10], the magnetic materials 5 (*e.g.* the ferromagnetic and antiferromagnetic materials) can have magnetizations (M) 10, which can be of different magnitudes and polarizations and which can induce the magnetic field effect in the graphene 1. The AFM can have layers 12 of magnetizations 11. Thus, a graphene magnet multilayer (GMM) can be formed, as shown in Fig. (**12a**) to (**12c**). (As is customary in the art, figures are not to scale and are drawn to illustrate the relationships between elements) [10].

In the structure plotted in Fig. (**12d**), the magnetic materials 5 in contact with the graphene 1 in the GMMs are discontinuous discrete sections 13 of magnetic materials 5. The sections 13 can be arranged as a patterned array, where each section has substantially identical magnetizations to achieve a desired electrical characteristic, such as, for example, Zeeman splitting. Alternatively, one or more sections 13 can have different magnetizations so as to achieve a desired electrical characteristic in a spintronic device, such as, for example, different spin-polarizations to facilitate spin-flipping of a spin-polarized electric current [10].

Strong hybridization of carbon sp^2 orbitals with d orbitals of magnetic 3D metals, either direct or through an intervening oxygen p orbital in magnetic oxides, exposes graphene electrons to an effective Weiss exchange field of enormous strength (a few hundred Kelvins or more) and can produce magnetic band splitting similar to that shown in Fig. (**11b**) and (**11c**) [10].

The use of GMMs allows creation of the full spectrum of devices which can be re-configured by appropriately changing the magnetization (M) 10 of the magnetic material(s) 5. Re-writable microchips and processors created on a single GMM can be implemented based on the GMMs described herein. In some embodiments, this change in magnetization can be facilitated by using temperature or light, as in the case of light-assisted (re-) magnetization [10].

Exchange fields experienced by electrons in a magnetic medium result from orbital overlaps and hybridization, such that electrons belonging to different atoms share common orbital states. Magnetic proximity effects arise from orbital hybridization at the interface between magnetic and non-magnetic materials, or between two magnetic materials. As a result, at the interface with magnetic material atomic electrons can experience the effective exchange field, which varies in strength among material pairs. Such exchange proximity effects are well documented in magnetic thin films and multilayers. For example, exchange coupling at the interface of the antiferromagnetic NiO or CoO whose bulk ordering temperatures (Neel temperature, T_N) are $T_N^{NiO} \approx 520$ K and $T_N^{CoO} \approx 291$ K, respectively, and the ferrimagnetic magnetite Fe_3O_4 (with a Curie temperature (T_C) of 858 K), induces antiferromagnetism in few-layer NiO and CoO films on Fe_3O_4 at a temperature one hundred or more degrees above T_N (up to $2T_N$ for CoO). Similarly, CoO antiferromagnetic ordering in NiO(2.1 nm)/CoO(1.5 nm) multilayers is shifted towards T_N^{NiO}, occurring about 100 K above T_N^{CoO}. All these proximity effects can be successfully understood on the mean field level as caused by an effective exchange field acting at the interface [10].

At the interface of graphite or graphene and a ferromagnetic or ferrimagnetic material, such as iron or magnetite, there may be a significant magnetic proximity effect. Angle-resolved photoemission spectroscopy (ARPES) of MG on a Ni(111) substrate indicates significant hybridization between graphene and the nickel conduction band. Obtaining magnetic band splitting in graphene by the magnetic proximity effect while avoiding distortion of graphene π bands through hybridization with the conduction band(s) of the substrate, may be accomplished by using insulating antiferromagnetic oxides, such as NiO and CoO, as described above. In these materials antiferromagnetic order exists in the form of ferromagnetic sheets with alternating magnetizations. Therefore, graphene on the uncompensated (111) surface may experience a homogeneous Weiss field by virtue of C—O—M (Carbon-Oxygen-Metal, where M=Ni, Co, Fe, for example) hybridization. It should be noted that the triangular symmetry of (111) planes in the cubic metal oxide rocksalt structure matches well with the hexagonal symmetry of graphene layer(s). By using solid solutions of AFM oxides such as $(Ni_{1-x}Co_x)_{1-y}(Al,Mg)_yO$ it may be possible to tune Zeeman band splitting in graphene induced by the effective Weiss (exchange) field by virtue of the proximity effect from 0 to about 0.05 eV. The orientation of this Weiss field from AFM metal oxides can be manipulated by changing the magnetization of the under/overlying ferro/ferrimagnetic layer such as Ni, Fe, or Fe_3O_4, again employing the magnetic proximity effect [10].

The crossing of the energy bands associated with two different sub-lattices in graphene results in the energy spectrum of electron and hole quasi-particles which is linear in momentum k (see Fig. (**11a**)): $\varepsilon(k)=v_Fk$, where $v_F \approx 10^6$ m/s is the Fermi velocity. Such quasi-particles are formally described by the Dirac equation for massless fermions. One consequence is the conservation of "chirality" of quasiparticles, defined as a projection of the pseudospin on the momentum, $k\cdot\sigma$ [10].

A charge density wave establishes different population densities on the two sub-lattices of graphene. A charge density wave may not form in an isolated graphene sheet, but its formation may be facilitated by an external magnetic field perpendicular to the graphene plane. The formation of a spin density wave may also be facilitated by an in-plane magnetic field [10].

As described with respect to Fig. (**11b**), an in-plane magnetic field effect based on the exchange magnetic field has two effects. First, it induces Zeeman splitting of the electron and hole bands such that these bands overlap and second, it leads to extended Fermi surfaces for electrons and holes with a finite density of states at zero energy. In such a situation, the Coulomb attraction between electrons and holes can lead to formation of a chiral excitonic condensate (CEC), the condensate formed by bound states of electrons and holes with opposite chiralities. In the presence of a CEC, the chiralities of electrons and holes in graphene are no longer conserved. The formation of a CEC leads to the emergence of a gap in the quasiparticle dispersion such that graphene becomes an excitonic insulator, Fig. (**11c**). By magnetizing the magnet layer, the conductivity of graphene in a graphene-magnet multilayer (GMM) may be manipulated [10].

The critical temperature for the disappearance of the excitonic gap 9 (Fig. (**11c**)) in the energy spectrum in graphene has been estimated as T_c ~0.1 B. This temperature may be significantly higher in double-layer graphene because of the larger overlap between the electron and hole bands caused by the quadratic gapless form of the quasi-particle spectrum in this material. When the strength of the effective Weiss magnetic field induced by the magnetic proximity effect in GMM is high, up to 10^3 K, then the excitonic insulator phase in GMM may be formed at room temperature [10].

The technology described by authors [10] exploits the spin-dependent splitting of the valence bands in graphene in a magnetic field induced in GMMs due to the magnetic proximity effect. In these artificially layered structures, conducting and magnetic properties are controlled by different materials and in different layers, so that they can be tuned and optimized independently. Samples of graphene display almost ballistic conduction under some conditions, exhibiting extremely low resistance due, presumably, to little or no scattering. Because of the ballistic conduction in graphene, the magnetic proximity effect may result in coherent spin-polarized transport in GMM. Specific embodiments of representative GMM spintronic nanodevices are described with reference to Figs. (**13**) to (**15**) [10].

Fig. (**13a**) to (**13c**) depict exemplary spintronic devices based on electric transport by polarized charge carriers in magnetized graphene. (It should be noted that these figures are drawn with right angles and terminals at edges of regions of interest for ease of illustration. Real world devices can be of any convenient shape and contacts need not be regular or symmetric, but need only make electrical contact with areas of the graphene sheet spaced some distance from each other.) Charge carriers can include negatively charged electrons and positively charged holes [10].

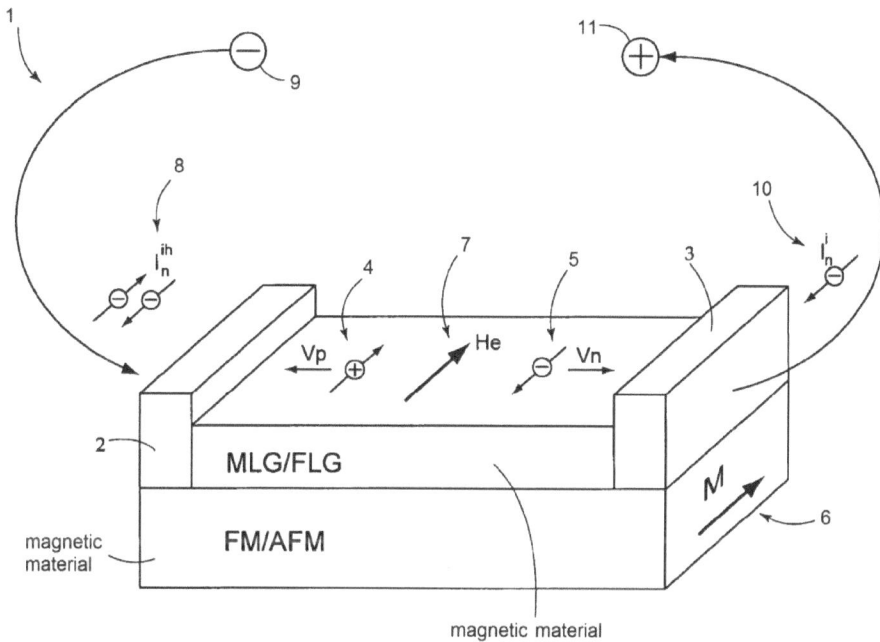

Figure 13a: [10]

Fig. (**13a**) shows a diagram of a spintronic polarizer 1. In an electric field obtained by applying a voltage to the electrodes 2 and/or 3, holes 4 and electrons 5 drift in opposite directions. In this embodiment, the magnetic material(s) have a magnetization (M) 6 and the graphene sheet experiences a magnetic field effect (H_e) 7 induced by the magnetization (M) 6 of the magnetic material(s). The magnetized graphene layer acts as a polarizer of electric current. If an unpolarized electric input current 8 is supplied by the "negative" terminal (source) 9, AP electrons 5 with antiparallel spin move to the right and will be transmitted *via* electrode 3 producing a polarized electric output current 10 at the positive terminal 11. Electrons with parallel spin polarization will recombine with holes 4 moving to the left [10].

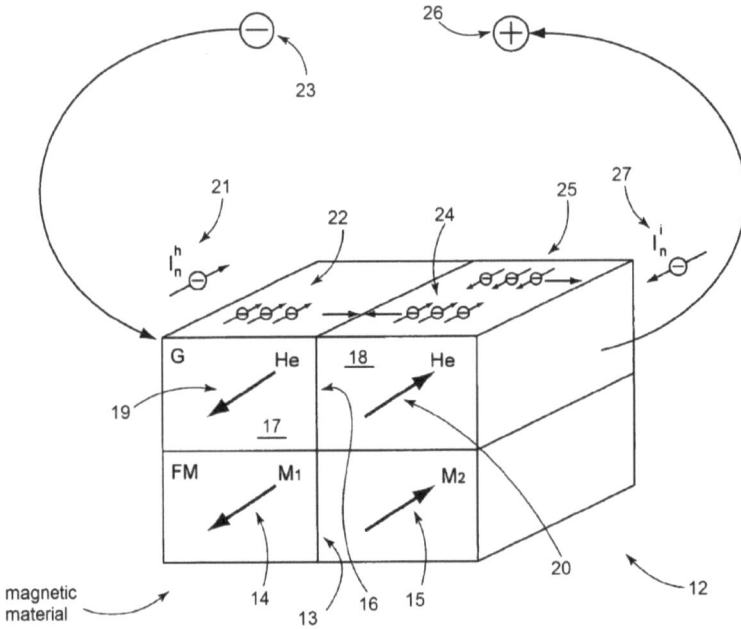

Figure 13b: [10]

When working with beams of spin-polarized particles a device which flips the particle spin is an essential element in the polarized beam setup. Such a spin flipper device 12 is depicted in Fig. (**13b**). Electrode terminals have been left off for clarity, but those skilled in the art will recognize that electrode terminals can be placed at each end of the graphene. The spin flipper device 12 can include graphene and magnetic material(s) having a domain wall 13 in which the magnetic material to the left of the wall has a first magnetization (M_1) 14 and the magnetic material to the right of the domain wall 13 has a second magnetization (M_2) 15. In the present example, the magnetizations 14 and 15 are in opposite directions. The domain wall 13 in the magnetic material(s) induces a wall 16 between two graphene domains 17 and 18, which are magnetized according to a magnetic field effect induced by the magnetic material so that the domains 17 and

18 are magnetized in opposite directions having magnetic field effects (H_e) 19 and 20, therefore acting as a spin inverter (flipper) for the appropriately polarized electron current [10].

A polarized electrical input current 21 can be supplied to the spin flipper device 12 *via* the negative terminal 23. Spin-polarized antiparallel electrons 22 (with respect to the induced magnetic field 19) related to the polarized current 21 can flow towards the domain wall 16 where the spin polarized antiparallel electrons 22 recombine with spin-polarized parallel holes 24 (with respect to the induced magnetic field 20). Antiparallel (AP) spin electrons 25 can flow towards the right to the positive terminal 26 producing an output polarized current 27 that is polarized opposite that of the input polarized current 21 (*i.e.* flipped polarization) [10].

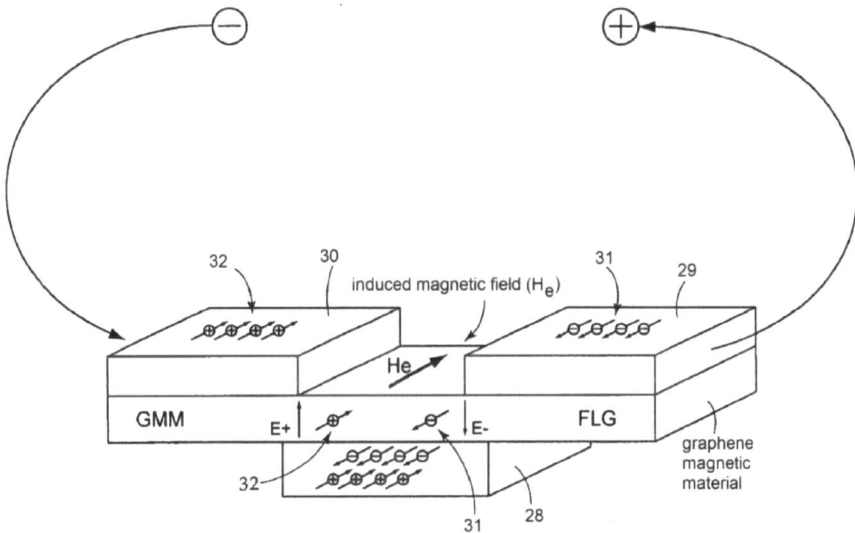

Figure 13c: Spin-polarized transistor [10] Fig. (**13a**) through (**13c**) depict exemplary spintronic devices based on electric transport by polarized charge carriers in magnetized graphene.

Fig. (**13c**) depicts a spin-polarized transistor having a GMM with an electrode terminal 28, an electrode terminal 29, and an electrode terminal 30. In the presented structure, the GMM includes a few-layer graphene (FLG) sheet having an induced magnetic field (H_e). The magnetic material(s) in the GMM have been omitted for clarity. Polarized electrons 31 and holes 32 moving in opposite directions in magnetic field (H) experience the same Lorentz force, *i.e.* polarized electrons and holes see opposite electric fields E_- and E_+. Hence, there is no transverse Hall effect. Net charge on the top surface will be zero, but AP spin electrons 31 and P spin holes 32 will be collected on electrodes terminals, 29 and 30, respectively. Applying a gate voltage to the gate electrode terminal 28, results in the spin-polarized transistor. In the magnetized GMM spintronic transistor shown in Fig. (**13c**) [10], the gate voltage is applied to the magnetized mono-/few-layer graphene film between the source and the drain leads. Such an electric field (supplied by the gate electrode shown on the bottom) destroys the symmetry between electron and hole Fermi surfaces, making one larger and the other smaller. Depending on the polarity of the applied voltage, this facilitates either electron or hole spin-polarized current and hinders the other. Thus, in this embodiment, the use of a magnetized graphene layer (*i.e.* graphene that experiences a magnetic field effect induced by an exchange magnetic field) produces a dependence of the current on the gate voltage similar to a field-effect transistor (FET), but for a spin-polarized current [10].

Fig. (**14a**) to (**14c**) depict magnetic tunnel junctions based on inhomogeneously magnetized graphene sheet in the GMM. In the present example, the FM and/or AFM materials as well as the electrode terminals have been omitted for clarity, however, those skilled in the art will recognize that the FM and/or AFM are part of the GMM as illustrated in Fig. (**12a**) to (**12d**) and that an electrode terminal can be disposed at each end of and in electrical contact with the graphene sheet. The graphene sheet in the GMM can contact differently magnetized magnetic material(s) resulting in different exchange magnetic fields in the graphene sheet. For example a first region 2 can have a first exchange magnetic field (H_{e1}) 3, a second region 4 can have a second exchange magnetic field (H_{e2}) 5, and a third region 6 can have a third exchange magnetic field (H_{e3}) 7. The different exchange magnetic fields can form domain walls between the regions 2, 4, and 6 [10].

In this example, exchange magnetic field (H_{e2}) 5 in the second region 4 is stronger than critical field of the excitonic transition (H_c), resulting in an excitonic insulator (EI) phase (Fig. (**11c**)), while exchange magnetic fields (H_{e1} and H_{e3}) 3 and 7 can be weaker than the critical field of the exictonic transition (H_c). Tunneling current depends on the thickness d of the EI segment (*i.e.* the second region 4) and on the relative orientation of the exchange magnetic fields 3, 5, and 7 [10].

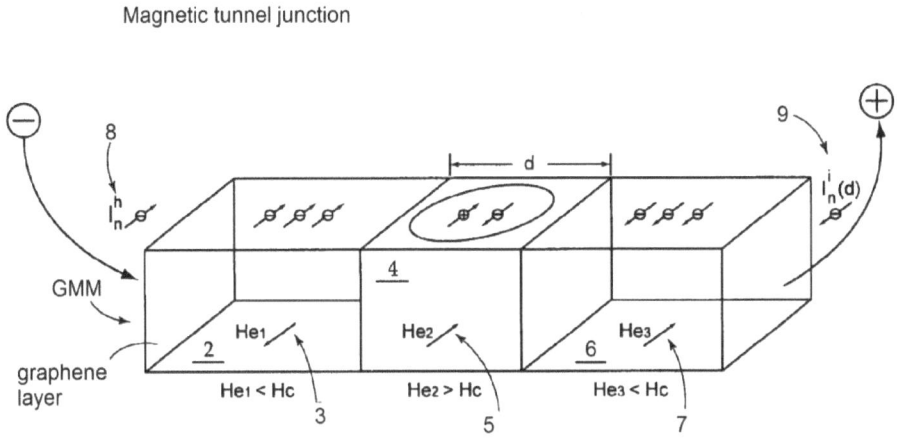

Figure 14a: [10]

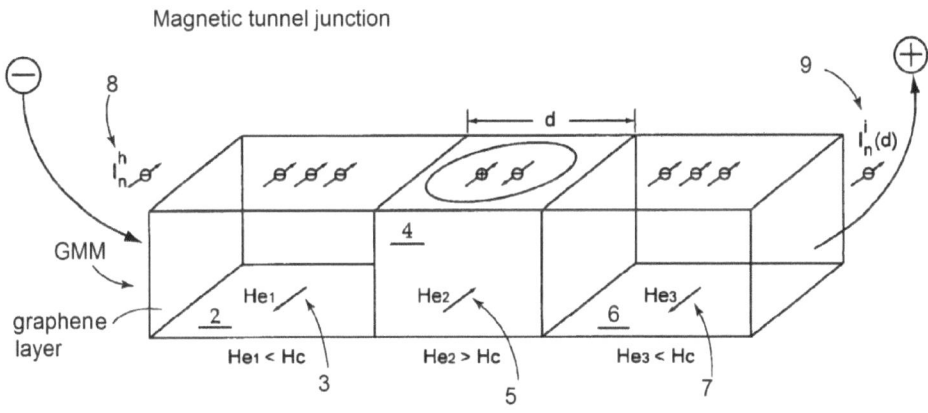

Figure 14b: [10]

Fig. (**14a**) depicts pair-assisted spin-flip tunneling through the excitonic insulator region (*i.e.* the second region 4). Tunneling electrons change spin polarization by virtue of breaking excitonic pairs. As shown in Fig. (**14b**) and (**14c**), non-spin-flip tunneling between graphene domains (*i.e.* regions 2, 4, and 6) in parallel fields is similar to tunneling through an ordinary insulator and is weaker than that depicted in Fig. (**14a**). Inhomogeneity in the magnetic field of the graphene may be induced by coupling a ferrimagnetic layer to it or by heating a small area of the ferromagnetic material with a light source and remagnetizing it with a different strength field. In addition, or alternatively, each region may use a different magnetic material having different magnetizations [10].

Opening of an excitonic gap in the EI phase in a strong magnetic field creates additional possibilities for manipulating the band structure of magnetized graphene. The excitonic phase is a small-gap semiconductor with spin-polarized bands and is sensitive to the combination of gate voltage and spin polarization of electric current. This can be utilized in an alternative scheme for the GMM transistor, where two leads are separated by the EI junction whose conductivity depends on spin polarization of the current and is controlled by the gate voltage applied to it [10].

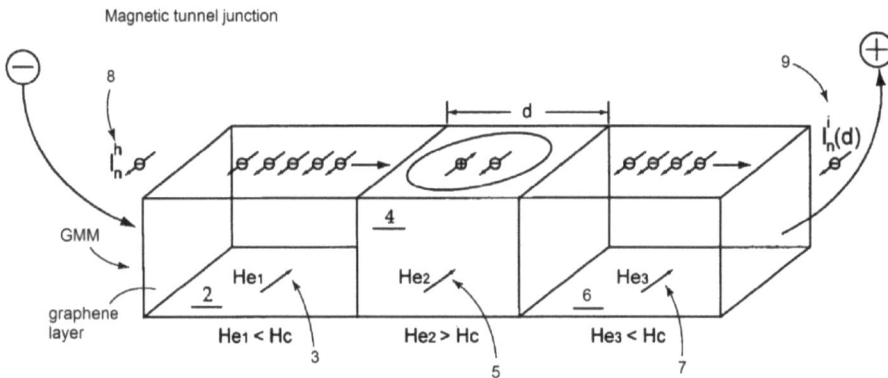

Figure 14c: **[10]** Fig. (**14a**) through (**14c**) depict magnetic tunnel junctions based on inhomogeneously magnetized graphene sheet in the GMM.

These devices can utilize the excitonic instability in graphene induced by a strong magnetic field effect described above. Such a field effect can be achieved where magnetization of the magnetic layer in GMM is sufficiently large, and transforms a part of the graphene layer into an excitonic insulator (EI). The magnitude of the tunneling current depends on the size of the EI region (*i.e.* the thickness d of region 4) and also on the relative orientation of magnetizations on both sides of the EI region (*i.e.* the first region 2 and the third region 6) and within the EI region (second region 4). A chiral excitonic condensate (CEC) is present in the EI region causing the polarization of charge carriers change orientation inside the EI region. The CEC is a bound state of electrons and holes of opposite ("left" and "right") chiralities, $e_L h_R + e_R h_L$. This phenomenon of chirality flip can be visualized in the following way: imagine, for example, a right-handed electron recombining with a right-handed hole from the condensate and thus liberating from it a left-handed electron [10].

Thus, in the structures depicted in Fig. (**14a**) to (**14c**) the spin polarized electric input current 8 supplied by the negative terminal can be reduced in magnitude, and in some active elements according to the invention [10]flipped in spin polarization, based on the thickness d of the second region and the relative orientation of the magnetizations of the regions 2, 4, and 6. For example, in Fig. (**14a**), the spin polarized electrical input current 8 is reduced and flipped to generate a smaller flipped spin polarized electrical output current 9, while in Fig. (**14b**) and (**14c**) the spin polarization is maintained at the output, but the magnitude of the current is reduced. [10]

Fig. (**15a**) and (**15b**) depict a spin-dependent quantum tunneling/interference device. In this structure, the FM and/or AFM materials as well as the electrode terminals have been omitted for clarity, however, those skilled in the art will recognize that the FM and/or AFM are part of the GMM as illustrated in Fig. (**12a**) to (**12d**) and that an electrode terminal can be disposed at each end of the GMM. The graphene sheet in the GMM can have differently magnetized magnetic material(s) resulting in different exchange magnetic fields in the graphene sheet. For example, a first region 2 can have a first exchange magnetic field (H_{e1}) 3, a second region 4 can have a second exchange magnetic field (H_{e2}) 5, and a third region 6 can have a third exchange magnetic field (H_{e3}) 7. The different exchange magnetic fields can form domain walls between the regions 2, 4, and 6 [10].

In the examples presented by authors [10] (Fig. (**15a**) and (**15b**)), the second region 4 is magnetized perpendicular (Fig. (**15a**)) or at an angle φ to the rest of the graphene sheet (Fig. (**15b**)). Both AP electrons 8 and P holes 9 injected into the second region 4 from the left and the right, respectively, experience spin precession around the field direction as they move. Quantum interference effects depending on the thickness d of the second region and the angle φ can lead to the recombination of electrons and holes in the second region 4, thus extinguishing electronic current [10].

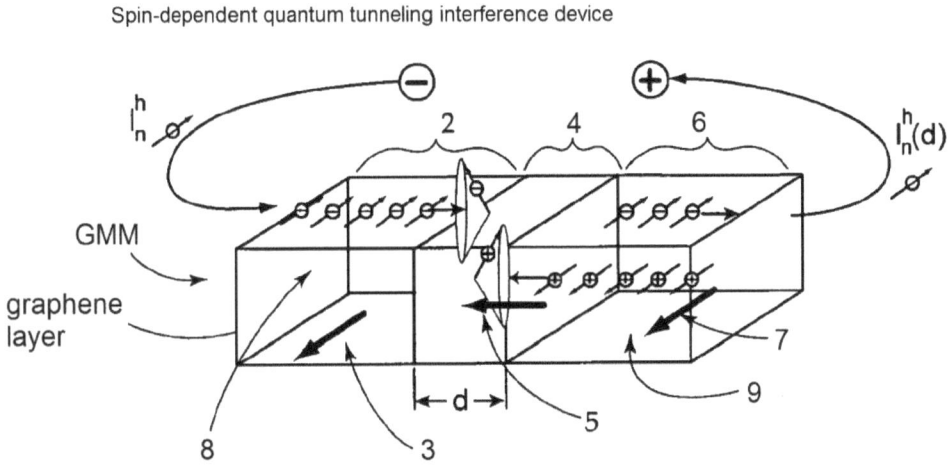

Figure 15a: [10]

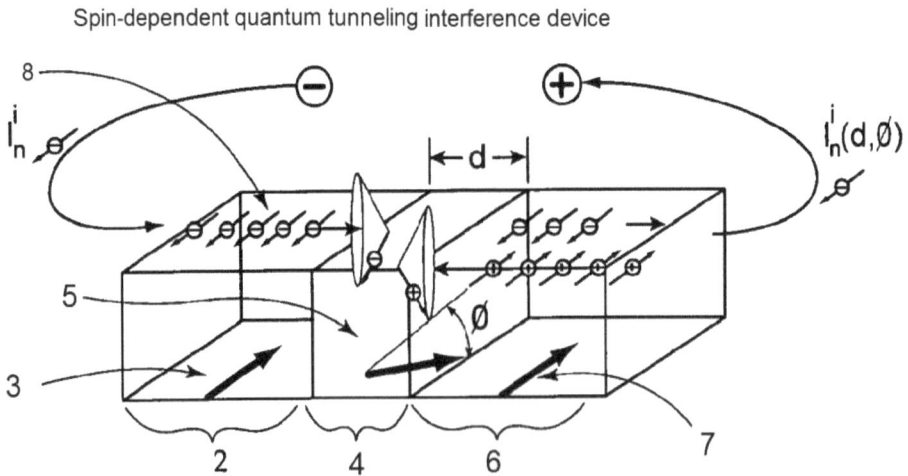

Figure 15b: Spin-dependent quantum tunneling interference device.

In these devices a region of the graphene sheet is subjected to a magnetic field oriented at an angle to the rest of the graphene sheet. The quantum transmission probability of the polarized current depends both on the size of the second region 5 and on the relative angle between the magnetizations [10].

3. OUTLOOK

As we can see, the spintronics proposes a variety of novel devices which open possibilities for non-traditional computing, data storage and information transfer unknown until recently. All interesting and important inventions of spintronics devices (only a few of them are presented in this lecture) are based on fundamental and experimental researches. It should be mentioned some achievements of special interest.

Nanotechnologies are widely used for designing future devices for quantum computation. Authors [13] have demonstrated a technique to nanofabricate nitrogen vacancy (NV) centers in diamond based on broad-beam nitrogen implantation through apertures in electron beam lithography resist. This method enables high-throughput nanofabrication of single NV centers on sub-100-nm length scales. Secondary ion mass spectroscopy measurements facilitate depth profiling of the implanted nitrogen to provide three-dimensional characterization of the NV center spatial distribution. Measurements of NV center coherence with on-chip coplanar waveguides suggest a pathway for incorporating this scalable nanofabrication technique in future quantum applications [13].

Nitrogen vacancies in diamond are appeared to be very interesting for potential spintronics applications. As noted in [14], the exceptional spin coherence of nitrogen-vacancy centers in diamond motivates their use in emerging quantum technologies. Traditionally, the spin state of individual centers is measured optically and destructively. Authors [14] demonstrate dispersive, single-spin coupling to light for both non-destructive spin measurement through the Faraday effect and coherent spin manipulation through the optical Stark effect. These interactions can enable the coherent exchange of quantum information between single nitrogen-vacancy spins and light, facilitating coherent measurement, control, and entanglement that is scalable over large distances [14].

Traditional semiconductor-based structures also have found new perspectives for spintronics. Authors [15] have investigated thin Fe_3O_4 films grown on GaAs substrates. The spin dependent properties of epitaxial Fe_3O_4 thin films on GaAs(001) are studied by the ferromagnetic proximity polarization (FPP) effect and magneto-optical Kerr effect (MOKE) [15]. Both FPP and MOKE show oscillations with respect to Fe_3O_4 film thickness, and the oscillations are large enough to induce repeated sign reversals. The oscillatory behavior is attributed to spin-polarized quantum well states forming in the Fe_3O_4 film. Quantum confinement of the t_{2g} states near the Fermi level provides an explanation for the similar thickness dependences of the FPP and MOKE oscillations [15].

Successfully developed molecular-beam epitaxy and other technologies of preparing high-quality thin films on solid state substrates made it possible to produce a variety of structures with properties unknown until recently. Heterostructures ferromagnet-semiconductor are of special importance for perspective techniques of spin manipulation. Authors [16] demonstrate methods to locally control the spin rotation of moving electrons in a GaAs channel. The Larmor frequency of optically injected spins is modulated when the spins are dragged through a region of spin- polarized nuclei created at a MnAs/GaAs interface. The effective field created by the nuclei is controlled either optically or electrically using the ferromagnetic proximity polarization effect. Spin rotation is also tuned by controlling the carrier traverse time through the polarized region. Coherent spin rotations of 5π rad during transport are demonstrated [16].

Thus, the spintronics is now a conventional field of novel and traditional microelectronics. No lecture can present all achievements in fundamental and experimental investigations in the field of spin manipulation.

Therefore, a lot of success to young scientists who will be working in this branch of novel science!

ACKNOWLEDGEMENTS

Declared none.

CONFLICT OF INTEREST

The author(s) confirm that this chapter content has no conflict of interest.

REFERENCES

[1]   Jansen, R. The spin-valve transistor: a review and outlook. *J. Phys. D.: Appl. Phys.*, **2003**, *36*, R289-R308.

[2]   Zutic, I.; Fabian, J.; Das Sarma, S. Spintronics: fundamentals and applications. *Reviews of Modern Physics*, **2004**, *76*, 323-410.

[3]   Kasai, H.; Nakanishi, H.; Kishi, T. Surface-spintronics device. US20060186433. **2006**.

[4]   Huang, X.; Halilov, S.; Yiptong, J.A.C.S.F.; Dukovski, I.; Hytha, M.; Mears, R. J. Spintronic devices with constrained spintronic dopant. US20080012004. **2008**.

[5]   Kaushal, S.; Sugishima, K.; Ganguly, S. Spintronic transistor. US20080017843. **2008**.

[6]   Ishida, S.; Mizutani, S. Spintronocs material and TMR device. US20080063557. **2008**.

[7]   Gould, C.; Schmidt, G.; Molenkamp, L. W. Spintronics components without non-magnetic interplayers. US20090114945. **2009**.

[8]   Liu, J.; Sandthu, G. Spin current generator for STF-MRAM or other spintronics applications. US201000080047. **2010**.

[9]   Ando, K.; Harii, K.; Sasage, K.; Saitoh, E. Method for changing spin relaxation, method for detecting spin current and spintronics device using spin relaxation. US20100097063. **2010**.

[10] Zaliznyak, I.; Tsvelik, A.; Kharzeev, D.; Nanodevices for spintronics and methods of using same. US20100109712. **2010**.

[11] Crawford, T. M.; Garzon, S. Y. Nanoscale spintronic chemical sensor. US20100140109. **2010**.

[12] Park, S. Y.; Jo, Y. Relaxation oscillator using spintronic device. US20100301957. **2010**.

[13] Toyli, D. M.; Weis, C. D.; Fuchs, G. D.; Schenkel, T.; Awschalom, D. D. Chip-Scale Nanofabrication of Single Spins and Spin Arrays in Diamond. *Nano Lett.*, **2010**, 10 (8), 3168–3172.

[14] Buckley, B. B.; Fuchs, G. D.; Bassett, L. C.; Awschalom D. D. Spin-Light Coherence for Single-Spin Measurement and Control in Diamond. *Science*, **2010**, 330 (6008), 1212-1215.

[15] Li, Y.; Han, W.; Swartz, A. G.; Pi, K.; Wong, J. J. I.; Mack, S.; Awschalom, D.D.; Kawakami, R. Polarization and Magneto-Optical Kerr Effect in Fe_3O_4 Thin Films on GaAs(001). *Phys. Rev. Lett.*, **2010**, 105, 167203-1-4.

[16] Nowakowski, M. E.; Fuchs, G. D.; Mack, S.; Samarth, N.; Awschalom, D. D. Spin Control of Drifting Electrons Using Local Nuclear Polarization in Ferromagnet-Semiconductor Heterostructures. *Phys. Rev. Lett.*, **2010**, 105, 137206-1-4.

INDEX

www.ingramcontent.com/pod-product-compliance
Lightning Source LLC
Chambersburg PA
CBHW041702210326
41598CB00007B/502